The EU Water Framework Directive

An Introduction

Peter A. Chave

Publishing

Published by IWA Publishing, Alliance House, 12 Caxton Street, London SW1H 0QS, UK
Telephone: +44 (0) 20 7654 5500; Fax: +44 (0) 20 7654 5555; Email: publications@iwap.co.uk
Web: **www.iwapublishing.com**

First published 2001
Reprinted 2002
© 2001 IWA Publishing

Printed by TJ International (Ltd), Padstow, Cornwall, UK

British Library Cataloguing in Publication Data
A CIP catalogue record for this book is available from the British Library

Library of Congress Cataloging- in-Publication Data
A catalog record for this book is available from the Library of Congress

ISBN: 1 900222 12 4

Contents

Preface

The EU Water Framework Directive, which was published in the Official Journal of the European Community on 22 December 2000, is probably the most significant legislative instrument in the water field to be introduced on an international basis for many years. It stems from concerns amongst the Member States of the European Union over the disparate ways in which water is currently protected within the Community and reflects the moves towards integrated environmental management outlined in the environmental action programmes of the Community. It took over 10 years to develop and, to the end, engendered intense scientific and political debate within the Community. The interpretation of its many provisions is even now subject to scientific discussion and it is likely to be many years before all its ramifications are understood.

All countries in the EU have developed water protection policies and laws, and the EU itself had enacted a large number of individual legislative instruments by the early 1990s. However, these directives had been largely developed piecemeal, to address specific problems. There was concern that groundwater was not adequately protected, both in terms of its quality and the ever increasing needs for water supply. It was further felt that there was a need to harmonise the quality of surface waters, particularly in respect of their essential role in safeguarding natural flora and fauna. Suggestions in the 1980s for legislation designed to protect aquatic ecology, and the partial development of a European policy for groundwater generated much discussion, and contributed to the development of the present Directive. The Commission produced a water policy in 1995 and followed this by publishing a proposal for a Water Framework Directive in 1997. It has taken the remaining years until now for the Directive to be finally agreed by the Member States. The difficulties of producing such an overall framework for water management can be judged by the complexity of the resulting document.

The Directive takes a broad view of water management and has as its key objectives the prevention of any further deterioration of water bodies, and the protection and enhancement of the status of aquatic ecosystems and associated wetlands. It aims to promote sustainable water consumption and will contribute to mitigating the effects of floods and droughts.

Water management policy as set out in the Directive is focussed on water as it flows through river basins to the sea, and its provisions apply to all waters – inland surface waters, groundwaters, transitional (estuarine) and coastal waters. An integrated approach is introduces for water quality and water quantity matters, and of surface and groundwater issues, and the Directive introduces a framework for water management based on river basins. Water is thus seen as a coherent whole. The overriding objective of the policy is the achievement of "good status" in all waters. This is defined in the Directive. This objective is subject to a number of exemptions under certain circumstances, and indeed higher standards must be applied to some waters.

The Directive is one of the most complex introduced by the Commission in that it covers a whole environmental sector, water, in one instrument. Its implementation entails not simply the application of new technical standards but a requirement to introduce a whole new regime of management, based on river basins, irrespective of existing administrative or, in the case of international rivers, national boundaries. Whilst there are examples of the river basin approach in many countries the application of the Directive's provisions may involve Member States in establishing entirely new organisations within, and even beyond, their territories and this represents a huge challenge to governments and existing water management authorities.

This book looks at the major Articles of the Directive and attempts to draw together the various issues. It sets out my own views of the sequences of actions that are required to implement the Directive. Some examples of existing practices that could be used, or indeed some that must change, are described. For maximum usefulness, the book should be read alongside a copy of the Directive – to have reproduced the document itself would have made the book far too long. I have to emphasise that the views expressed are my own and do not represent those of any official body.

I would like to thank John Tyson for some enjoyable discussions on early versions of the Directive during work on a project in Hungary, and for suggesting (probably as a result of these) that I should write the book in the first place. Dr Peter Bird of the UK's Environment Agency answered many awkward questions, and provided much extremely useful information, for which I thank him. Thanks go also to Dr Alistair Ferguson of the Environment Agency for pointing me in the direction of information about the origins of the Directive, participants at the IWA Edinburgh Conference in January 2001, several of whom supplied information to me, Dennis McChesney of the US EPA for information and contacts at the Delaware River Authority. I thank also Dr Ingrid Chorus and a number of colleagues with whom I was involved in seminars on water pollution and groundwater protection on behalf of WHO, at the Federal Environmental Agency, Bad Elster who produced a number of useful ideas. I am also grateful to Vanessa Humphreys, Alan Peterson and Alan Click of IWA Publishing for their support in producing the book. Finally I thank my wife, Margaret, for giving me her encouragement and support during its writing.

Peter Chave
September 2001

1

Introduction

BACKGROUND TO THE DIRECTIVE

The development of water policy in the European Union

European Union (EU) policy in the water sector has largely developed through the political decisions taken in a series of five Environmental Action Programmes extending over the period 1973–2000, which will be followed in due course by a further programme currently being considered. The Action Programmes identified a number of priority issues that needed to be tackled to reduce water pollution and to improve the quality of natural waters in the countries of the EU. These included:

The definition of quality objectives. This concept was first adopted in the 1973 Action Programme as the identification of a set of requirements that must be fulfilled at a given time, in order that water may be used for a particular purpose such as bathing, drinking, or for the protection of aquatic life.
The control of dangerous substances. The early action plans took the view that is was important to give the greatest priority to the control of substances which

were known to be toxic, persistent and which bio-accumulate in the water environment, so as to reduce their polluting impact.

The protection of the sea against pollution. *Marine pollution was perceived as one of the most dangerous environmental problems because of the effect on the fundamental biological and ecological balances of the planet and this deserved action at Community level. This view is even more prevalent today.*

The surveillance and monitoring of water quality. *Monitoring of water quality was carried out in a wide variety of ways with little coordination of programmes or inter-comparability of methods. With the preponderance of rivers and seas that crossed international borders within Europe, monitoring in a form compatible between countries was seen as an essential background to pollution control making international collaboration a priority.*

The adoption of industry specific measures. *Many cases of pollution were caused by specific types of industry that were much more significant in the context of their impact on water quality than others. Protection of the environment required particular attention to be given to those industries that introduce pollutants into the environment.*

The development of international agreements on environmental protection measures. *Because the solutions to many water quality issues relied upon the efforts of more than one country acting alone where waters – rivers or the sea – crossed country borders, international agreements and protocols are essential in setting up and maintaining measures to improve water quality in the sea and international river systems.*

The funding of research and development work. *It was considered essential to ensure that knowledge and understanding of the impact of pollutants on the environment was gained as an essential step in solving problems.*

Early European legislation was therefore aimed at dealing with these issues and resulted in a large number of directives dealing with specific issues. For example, the Bathing Water Directive (76/160/EEC) (EU 1976a) sought to improve the quality of natural waters used for bathing by specifying quality limits on contaminant levels, and the Dangerous Substances Directive (76/464/EEC) (EU 1976b) was aimed at the control of the discharge of the most dangerous chemicals into the water environment by limiting the discharge of individual substances. Over the same period a number of international conventions and protocols were developed and ratified covering such rivers as the Rhine, and seas such as the Mediterranean and the North Sea, and more general international collaboration on protection of the marine environment through the Paris Convention of 1974 (Paris 1974).

Approaches to pollution control

Throughout the development of policy for the control of water pollution and the improvement of water quality, there has been debate over the most effective way of exercising control over the causes of pollution.

Point sources of pollution, that is, generally, pollution arising from the discharge of materials through a defined orifice (a pipe or channel) from industrial plants, or urban waste water treatment works, can be controlled by applying some form of regulatory standard to the quality of the effluent that is discharged. Commonly, national standards for particular polluting materials have been determined, and a permitting regime established, within which such discharges are authorised and the conditions placed upon the discharge reflect the national standards. This is the approach that prevails in many of the countries comprising the European Community, and EU legal instruments simply superimpose prescribed Community emission standards in place of or in addition to national limits. Whilst this approach is relatively straightforward to apply, the resulting emission standards do not always take account of the effect of the discharge on the local water environment, with the result that in some cases the standards may be unnecessarily stringent and in others too lax. An alternative approach preferred by some Member States has been to identify the uses to which the receiving watercourses are to be put, and to identify the required degree of purity of the water for this purpose. The resulting water quality standards, applied to the water rather than the effluent, become a set of goals at which pollution control measures are aimed. The measures then taken to meet these water quality objectives are not restricted to the application of emission limits, as a means of control, but may involve other methods such as regulation of the use of raw materials or specifying the parameters of operating industrial processes. Of course limit values are also an alternative, but limits placed upon individual discharges are then tailored to meet the water quality objectives. Such limits may vary from place to place even when applying to identical industrial processes, but the limit values take into account such factors as the amount of dilution available or the self-purification properties of the water.

In the European context these two approaches have become known as the Water Quality Objective (WQO) approach and the Emission Limit Value (ELV) approach. Although they are different in concept the two systems of control are not incompatible when applied to point source discharges. However, where there are no identifiable specific entry points for pollutants, the ELV approach has limited applicability, whereas it is possible to use the WQO method to set the required standards and rely upon other means of control to achieve those values.

The ELV approach appears relatively straightforward to apply and to enforce. It also enables common operating standards to be required for similar types of industrial plant throughout the Community. On the other hand, the application of emission standards is only feasible where there are identifiable discharges. Diffuse pollution from land run-off and industrial practices which do not give rise to an identifiable discharge cannot be controlled in this way. The use of WQOs may, on the other hand, give a target to aim for in the receiving watercourse, and anti-pollution measures of an entirely different kind, such as controls on manufacture, distribution and use of potentially polluting materials may be adopted, and the effectiveness of the controls may be judged against achievement of a measurable quality of water.

Both approaches have their merits, and in practice an environmental policy based on a "parallel approach" has developed. Some of the directives within the water sector require the setting of emission standards for effluents that are discharged to the environment; others require certain quality standards for the water to be met to enable it to be used for its specified purpose, for example bathing or drinking; whilst other directives contain both alternatives and leave the choice of method to Member States, (for example, the daughter directives of the Dangerous Substances Directive (76/464/EEC)). Whichever method is used, most of the directives specify that a system of prior authorisation by the regulatory authorities is instituted so that discharges are controlled through a permit. The application of either the WQO or the ELV approach means that any numerical standards, which apply to the discharges, are derived as either national fixed emission standards or standards set at such a level as to ensure that the receiving water meets its own quality standards.

In the past, Member States have tended to more frequent use of the use of fixed emission standards. However, the benefits of formulating targets for the long-term state of the environment in order to facilitate planning has recently become more accepted, and this has resulted in a move towards a greater use of the quality objective approach in the EU Water Framework Directive, where river basin plans will require the adoption of measure to achieve specified objectives.

Recognising the limitations of using the approach of taking action on individual problems by specific regulatory instruments the 4th Environmental Action Programme attempted to consolidate efforts to improve the environment by emphasising the need to implement existing Community legislation and to regulate environmental impact of substances and sources of pollution. The 5th Environmental Action Programme took a different approach by looking towards the long-term sustainable management of natural resources: soil, water, nature reserves and coastal areas. In the water sector this entails taking an overall, integrated view of the quantity and quality of water which is available, how it is utilised, and what measures are needed to protect it over the long term. The 6th Environmental Action Programme is expected to concentrate on the concept of sustainability.

EUROPEAN WATER POLICY

Water policy took a step forward in 1995 when acting on a request of the Council of Ministers and the European Parliament the European Commission put forward a view on water policy that took account of the overall community environmental policies set out in Article 130 of the Treaty of Rome. The new principles included the following issues:

Adoption of a high level of protection – *in the context of water management this means ensuring that water resources and their associated ecosystems receive protection against pollution and deterioration at a higher level than the minimum attainable.*

Adopt the precautionary principle – *even in the absence of conclusive evidence of a deleterious effect caused by certain substances or activities on the water environment, a precautionary approach to their use should be adopted.*

Take preventative action – *pollution of the water environment should be avoided by taking steps to prevent the entry of polluting matter in order to avoid potentially long term and expensive remediation.*

Deal with pollution at source – *if a polluting situation is identified, the source of pollution should be addressed rather than dealing with the problem at a later stage.*

Adopt the polluter pays principle – *the polluter should in general bear the cost of pollution prevention, control and remediation.*

Integrate water and environmental protection as a part of other sectoral policies – *as activities in many other sectors can influence water quality and quantity, environmental policies should become integrated into other sectors such as land use planning, transport and agriculture.*

Use available scientific and technical data – *to make informed decisions in the water sector and apply the best available techniques for prevention and treatment of environmental problems.*

Take account of the variability of environmental conditions in the regions of the Community – *an acknowledgement of the need for flexibility to avoid the imposition of inappropriate or unnecessarily strict standards and allow for regionally specific problems.*

Take account of costs and benefits – *the development of a cost effective strategy through the use of regulations and standards, new technology and pricing and market based incentives.*

See water policy as a contributory element of a balanced and sustainable economy.

Recognise the need for international collaboration.

Adopt the principle of subsidiarity – *measures that can be undertaken more effectively at MemberState level should not be undertaken at Community level.*

When compared with these new guiding principles, the existing environmental legislation was seen to be rather insular, with directives dealing with individual problems without necessarily relating to the whole water environment. A number of other initiatives were under way that further reinforced the possible need to revise the way in which the water environment was being managed at Community level.

A resolution of 28 June 1988 (EU 1988) specifically required action to improve ecological quality of surface waters in the Community and a conference at Como in May 1989 (Como 1989) concluded that it would be possible to prepare a proposal that would cover the ecological quality of all Community waters, complementing, but not replacing, existing directives in the water quality area. The conference proposed fundamental aims to attain high ecological quality in surface waters; and that Member States would set up targets in terms of quality levels for their aquatic systems, with timetables for their achievement water; Guidelines for this would be laid down in the new directive. Classification schemes for quality levels, based at first upon existing schemes within Member States, would be harmonised. Provisions for monitoring and control, and public accessibility would also have to be agreed, and a comprehensive reporting and publication of information showing improvements in water quality would be features of the new system. This conference concentrated upon the ecological quality of surface waters, but at the same time concern was gathering over the state of groundwater quality and quantity in the EU.

A Council Resolution on groundwater policy of 25 February 1992 (EU 1992) that underlined the vital importance of groundwater for human health and for all forms of life and ecosystems, expressed concern at falling groundwater levels and rising pollution of aquifers and called for an action programme to improve the situation. In 1996 a proposal for an action programme for the integrated protection and management of groundwater was presented which drew attention to the need for the regulation of the abstraction of groundwaters and to links with the monitoring of freshwater quality and quantity (EU 1996). A number of other official decisions and resolutions had a formative role in the development of the directive including the declaration of the Ministerial Conference on groundwater at The Hague in 1991 (EU 1991) and the results of the European Environmental Agency's (EEA's) state of the environment report (EEA 1995).

As a result the Commission consulted upon the potential for a more integrated approach to water management and in February 1997 published a proposal for a new framework directive for the water field. This adopted a holistic approach to water management, taking into account existing legislation in the water field and also relevant legislation from other environmental sectors that impinge upon water quality and quantity. The proposal took into account the suggestions made in a draft directive on the ecological quality of water that had been put forward in June 1994, which itself had been aimed at meeting the request in the 5th Environmental Action Programme for measures to improve the ecological quality of surface waters (EU 1993). A further three years of negotiation and discussion passed before the final version of the directive was approved by the Conciliation Committee of the Council, Parliament and Commission under the principle of co-decision on 30 June 2000 and published in the Official Journal on 22 December 2000. The final version of the Directive is remarkably similar in form to the original proposals put forward at the 1989 conference. These suggested that national systems of water quality classification should be translated into a pan-European system and plans should be made to reach, eventually, a common level of ecological quality. The proposal envisaged a number of steps that involved identifying the waters, describing their present ecological quality, identifying the sources of pollution, setting desirable levels of quality and implementing an action programme to meet these levels over a realistic timetable.

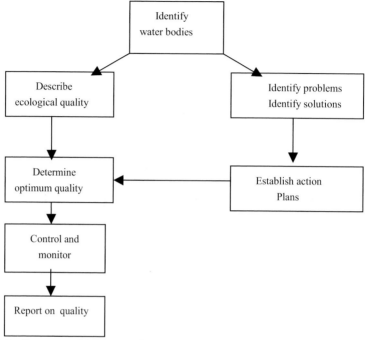

Figure 1.1 Outline proposal for an ecological quality directive

REFERENCES

Como (1989) The ecological quality of surface water: preparation of a Community Directive. European Commission.

EEA (1995) Report on Environment in the European Union, European Environment Agency, Copenhagen, 1995.

EU (1976a) Council Directive 76/160/EEC concerning the quality of bathing water, OJ L131, 5.2.76.

EU (1976b) Council Directive 76/464/EEC on pollution caused by certain dangerous substances discharge into the aquatic environment of the community, OJ L129, 18.5.76.

EU (1988) Resolution concerning proposals for improvement of ecological quality in surface waters, OJ C 209, 9.8.88.

EU (1991) Declaration of the Ministerial Conference on groundwater, The Hague.

EU (1992) Council Resolution on groundwater policy, OJ C 59/2, 6.3.92.

EU (1993) OJ C 138, 17.5.93.

EU (1996) Proposal for a European Parliament and Council decision on an action programme for integrated groundwater protection and management OJ C 355, 25.11.96.

Paris (1974) Convention for the prevention of marine pollution from land based sources.

2

The EU Water Framework Directive

POLICY AND PRACTICE

How does the new Directive meet the policy provisions described in Chapter 1? To answer this question, reference to the preamble to the Directive is necessary. As with all similar directives, the thinking of the Commission is set out as a number of statements that describe the principles upon which the Directive is based and give an insight into the deliberations of the Commission and the governments of the Member States.

The statements confirm that the Commission believes that an integrated policy on water is necessary and that there is need for close cooperation at all administrative levels and with the public. The policy provisions of preventive action, the precautionary principle, and the polluter pays principle are adopted and the causes of environmental damage should be rectified at source. Local situations are to be taken into account by the ability to adopt different specific solutions taking into account the diversity of conditions throughout Member States with local responsibility for action. The Directive is seen as making a significant contribution to cooperation between Member States through integration with other sectoral policies. There is specific mention of the impact that inland surface water may have on the coastal zone with improvements influencing the economics of, for example, coastal fisheries.

The Directive is seen by the Commission as a providing a framework for each country to develop a common basis for the protection and sustainable use of water. It's overall aim is to maintain and improve the aquatic environment through attention to quality issues, but incorporating the control of quantity as an essential ingredient, recognising the impact that inadequate quantity could have on the maintenance of good ecological quality.

The European Community is a participant in a number of international conventions that relate to the prevention of pollution in the marine environment, for example the Helsinki Convention (Helsinki 1994), the Paris Convention (Paris 1998) and the Barcelona Convention (Barcelona 1977). The directive aims to assist in meeting these conventions by focussing attention on the reduction of the discharge of certain priority dangerous substances in water to such an extent that ultimately there should be a concentration of these substances no higher than natural levels in marine waters.

OBJECTIVES OF THE DIRECTIVE

In addition to establishing a new, common management system for the delivery of water policy, the EU Water Framework Directive contains a set of overall objectives. These are:

- expand the scope of actions to protect water to all forms of naturally occurring water in the environment, including surface and groundwater;
- prevent further deterioration, and protect and enhance the status of aquatic ecosystems, and with regards to their water needs, terrestrial ecosystems and wetlands (Article 1(a));
- promote sustainable water use based on long-term protection of available water resources (Article 1(b));
- take specific pollution control measures, by reducing or eliminating discharges and emissions and losses of priority toxic substance, to enhance the protection and improvement of the aquatic environment (Article 1(c));
- reduce pollution of groundwater (Article 1(d));
- contribute to mitigating the effects of floods and droughts(Article 1(e)).
- undertake measures which will result in the achievement of 'good water status' for all waters within a predetermined timescale.

As a result of these objectives, the Commission expects to maintain a sustainable balanced and equitable use of water in the countries of the

Community; to reduce groundwater pollution; to protect territorial and marine waters; and meet international obligations related to toxic substances.

BASIC FEATURES OF THE EU WATER FRAMEWORK DIRECTIVE

Perhaps the four most important and innovative features of the Directive are that it aims:

- to manage water as a whole on a river basin basis reflecting the situation in the natural environment;
- to use a combined approach for the control of pollution, setting emission limit values and water quality objectives;
- to ensure that the user bears the costs of providing and using water reflecting its true costs; and
- to involve the public in making decisions on water management.

In order to do this a number of issues will have to be addressed, some of which are used already in certain countries, others of which are new.

River basin management

The new approach recognises that water, in its natural environment, is generally related in some way to river systems. In the natural world, water naturally drains from the upper reaches of a watershed towards a river network. Although diversions such as man-made reservoirs, natural lakes and canal systems may interrupt the drainage pattern, eventually the water reaches those related environments of estuaries and the near-coast sea. Underground water in aquifers within the catchment areas, although constrained and diverted by the geological strata, is also related in many ways to the natural boundaries of rivers, outcropping at springs and often affecting the base flow of river systems. Although the boundaries of aquifers do not always match river basin boundaries, it is reasonable to relate aquifers to their nearest or most practical river basin. The Directive therefore adopts the river basin as the natural management unit for the protection of water.

In order to manage the water environment as a whole, the Directive requires Member States to identify river basins and all of their associated surface and underground waters. The size of the basins must be sufficiently large to permit the establishment of an organisation to assume responsibility for their administration. In order to achieve this, small river basins may be combined (or excessively large ones sub-divided). The Directive encompasses river basins that cross international frontiers so any organisations set up must be capable of dealing with the international negotiations necessary to achieve the objectives.

In this respect this Directive is likely to be one of the most significant legal instruments yet adopted in the environmental field as it directs how an environmental sector is to be managed, institutionally, and as a whole.

River basin plan

The fundamental tool of river basin management, which will be used by the relevant institutions, is the preparation of a river basin plan. The Directive requires such a plan to be drawn up on the basis of full knowledge of the actual situation in each river basin in terms of the condition of the water and those factors which may influence it. This information will be acquired through a comprehensive survey carried out early in the life of the Directive, which will examine the physical and geographical characteristics, industrial activities, populations and their activities of the river basin, a review of the environmental impact of human activity, and an economic analysis of water usage. The plan will take account of the current status of water throughout the river basin, comparing it with criteria that define the status of the water in the river basin, both in terms of quality and quantity. Once the actual status has been determined, environmental objectives will be set for the ecological and chemical quality of surface water and the quantitative and chemical status of groundwater. The aim of the plan is to achieve the objectives, and the Directive specifies that the objectives should usually be a condition known as "good" status. Good status for individual water bodies will depend on their type and their geographical location in Europe, and is identified by a number of ecological and chemical criteria. The plan will identify shortfalls in status when compared with the desired "good" status, and measures must be taken which will be aimed at achieving this status generally within 15 years from the date the Directive comes into force. The Directive allows some particular exceptions from this aim.

Programme of measures

The main means that will be used to achieve good status is the preparation and implementation of a "programme of measures". The programme will identify what needs to be done in terms of implementing existing legislation, and any other actions that are capable of being carried out by the competent body and others to bring the water up to the required standard within a timetable laid down by the Directive. Again this is seen as a significant step forward in that the Directive brings together many other legislative instruments in the water field, and also some from other environmental sectors, which are not normally associated directly with the water sector, in order to achieve the desired objectives. For example, directives in the nature protection sector, such as the Habitats Directive (92/43/EEC); those in the waste sector, such as the Landfill Directive (99/31/EC); and, most importantly, the Integrated Pollution Prevention and Control Directive (96/61/EC) which applies to the industrial sector. Earlier, but important, directives relating to the assessment of risks and control of chemicals are also brought into the framework through a requirement for action by the Commission to prioritise certain substances for control through the procedures of Directive 98/8/EEC and other instruments. For such priority substances Community emission levels and water quality objectives have to be established.

Consultation

Because the plan may affect the lives of people who live in the river basin districts through, for example, the adoption of stricter controls on their activities, or the need to pay for remediation or improvement work, extensive consultation is anticipated. The consultation process is specified in the Directive through responsibilities placed on the competent authorities to publish plans, to provide information, and to consult with the local population and with other interested parties within specified time scales, so that ample opportunity is given for them to make their views known before any plans are adopted. The Directive also lays down procedures for recording and reporting the results of the activities that are undertaken through the programme of measures and information on the state of the aquatic environment from the monitoring programmes that are established to assess their success or otherwise. Member States have reporting deadlines to meet to provide the Commission with information, and the Commission itself has a heavy reporting schedule to provide information to the European Parliament and the Council of Ministers on how the Directive is being implemented in Member States. Most of the reporting will be undertaken through the medium of the river basin plans.

Combined approach

As explained earlier, the development of mechanisms for pollution control and the prevention of environmental damage have followed two distinct courses — the control of emissions, and the establishment of standards applicable to the water body. Whilst in a few cases these operated together, in the majority of legislation they were separated and often operated in parallel. The new approach in the EU Water Framework Directive is known as the "combined approach". In this case both the limitation of pollution by controlling the source and setting targets for quality to be met in water bodies will apply. Thus it will be expected that industries which discharge polluting materials into watercourses will be subject to controls that will limit their emissions from point sources, whilst at the same time the competent authorities will set numerical standards for chemicals in the receiving watercourses. These will reflect the required ecological status of the waters and provide upper limits of concentration for pollutants discharged from point sources together with any such pollutants that also gain access to the water from diffuse sources. This means that in a situation where both point and non-point sources are influencing water quality, actions to introduce a permitting regime for the control of entries from, for example, pipelines, together with other types of regulatory action for the control of diffuse pollution, may be required. This could, potentially, extend the role of the competent body significantly, possibly into areas unconnected with the water environment. At the very least, good communication will have to be established between the river basin district authority and any other regulatory bodies that may have controls over these other activities.

To enable such a wide range of activities to be undertaken, the competent authorities in each river basin district will need a high degree of legal power and large resources for they will have to identify all the various activities, discharges and other issues causing effects on water status throughout the river basin district. They will also have to be able to grant and enforce the conditions placed in permits for a wide range of activities and discharges; and to control other activities which may have adverse effects on water status.

Water quantity

In line with the holistic approach to water management outlined in the water policy, this Directive is the first to link water quantity with water quality. It is an acknowledged fact that the amount of water available at a particular point is a fundamental parameter in determining the concentrations of polluting chemicals, as a simple dilution factor is built into any permit calculations for emission limits. What has not often been taken into account previously is the effects that quantity can have on the ecological status of river stretches, or the ecology of lakes and reservoirs. Many species are influenced by the flow regime in rivers, and low (or high) levels in lakes can have significant influences on plant and animal populations. Not only this, because, in some cases, under drought conditions the groundwater associated with or flowing under a river system can become the dominant contributor to river flows, and thereby influence the availability of water to sustain the ecology and its quality, the quantity of groundwater becomes a factor in sustaining good water status. This Directive therefore links quantity with the attainment of good water status, and provides a basis for the control of abstraction, usage and recharge of water to aquifers. In the case of surface water the abstractions and return of water through the discharge of effluent will also have to be taken into account when permit procedures are invoked. Suitable plans will have to be drawn up to assess future water demand, including measures to limit this where necessary.

Cost recovery

A novel aspect of the Directive is the use of economic instruments to pay for water management and infrastructure development. The Directive requires the adoption of the "polluter pays principle" in respect of the discharge of polluting materials, but it also goes much further in that charges for using water must be brought in that reflect the true cost of its use. Water is abstracted and used for drinking, irrigation, and by industry, and watercourses are used as carriers of effluent and for leisure or navigational purposes; they also have an ecological and wildlife value. The Directive requires that the true economic value of water is taken into account and that full cost recovery is applied. This will be a radical change to the previous situation in many countries, and in most countries the costs for such uses as leisure are not normally fully recovered, except perhaps through general taxation or as part of the overall charges paid for water or sewerage services. There is a danger in this proposal that water may become too expensive a commodity for many, and a general reduction in its beneficial use may result. The Directive does indicate that the difficulties such increased charges may cause should be taken into account, particularly in respect of the provision of an adequate drinking water supply.

Monitoring

The Directive sets down a number of monitoring obligations. This will cover surface water status, groundwater status and the status of protected areas. The monitoring programme covers three aspects of water management:

- surveillance monitoring, where monitoring is undertaken to provide information on the status of water, to identify its initial condition and to assess long-term changes brought about by natural and anthropogenic activity;
- operational monitoring to assess the success or otherwise of measures enacted to improve the situation; and

- investigative monitoring following accidental pollution or to identify the causes of a problem, perhaps noted during other monitoring activities.

The results of all this monitoring must be made available to the public, the Commission and the European Environment Agency. Member States will be required to set up programmes and use approved, comparable methodology for their analysis.

DEFINITIONS

The Directive is a comprehensive attempt to integrate a diverse and complex area of environmental management. Although ostensibly aimed at a particular environmental sector, the water sector, it ranges into such issues as management and institutional structures, geographical and physical information systems, economics, physical, chemical, and biological sciences, and public communications. Covering the whole of the European Community area and beyond in so far as international river basins stretch beyond the borders of the existing and future EU, it is necessary to have some very clear definitions of many of the features to which reference is made in the Directive. Article 2 of the directive is entitled "Definitions" and runs to 41 Sections. This, to some extent, reflects the complexities of the issues at stake. Whilst some of the definitions are simple and included for clarity, others relate to much more complex issues and refer to further more detailed information set out in Annexes to the Directive. Even the simple definitions must be treated with respect as it is sometimes possible to interpret the specifications in different ways.

Definitions of water to which the Directive applies

A key element of the operation of the Directive is that it focuses water policy and activities on water in individual river basins as it flows towards the sea. Articles 1(1) to 1(12) define those waters which occur within such basins, and some waters beyond the boundaries, which should be included in any activities related to the implementation of the Directive. The definitions include surface water, groundwater, inland water, rivers, lakes, transitional waters, coastal waters, and aquifers. It is interesting and important to note that waters that are covered by the Directive include within the definition of inland waters *all groundwater on the landward side of the baseline from which the breadth of territorial waters is measured*. There is thus an imaginary cut-off of responsibility for groundwater quality where an aquifer extends under the sea bed. Not usually included in river basin considerations is the area of coastal water extending up to one nautical mile from the shoreline (technically the baseline from which the breadth of territorial waters is measured) or the outer limit of transitional waters. The latter are bodies of partly saline surface water in the vicinity of river mouths that are influenced by freshwater flows.

In terms of impact, countries within the EU use a variety of definitions for waters that come within their pollution control and water quality legislation. The new Directive will require adjustments to be made, and in some cases will add new water management dimensions to existing practices. For example the UK's water laws apply to "controlled waters" that extend to the limit of territorial waters, well beyond that specified in the Directive.

Water *bodies* (Art 2(10)) are discrete and distinct elements of water, such as a lake or part of a river. There are two important aspects of the definition that will require attention. Whilst it is a relatively simple matter to designate a lake or perhaps a whole river or tributary as a water body, it is not such a simple matter to differentiate between the limits of coastal water bodies in such a way that their characteristics vary. Commonly, such delineations are simply drawn on a map and are "administrative boundaries". The Directive requires water bodies to be distinct entities that can be described by their scientific or physical characteristics so that their ecological quality can be ascertained and then monitored. At present the limits of estuaries or bathing waters, or even stretches of major rivers, are delineated largely for administrative convenience. The Directive will require more scientifically justified boundaries to be drawn. Annex II sets out descriptors for the various types of water body and reference to this will be required to enable a formal identification to be made in some waters.

It should be noted that groundwater is defined as water in the saturated zone rather than all groundwater, and underground water may be further defined as a *body of groundwater* – a distinct volume of groundwater within an aquifer.

Two definitions are important in the context of water status. "Artificial water body" (Article 2(8)) means a body of *surface* water created by human activity (such as a dewpond or reservoir or perhaps even an effluent treatment lake of significant size); and "heavily modified water body" (Article 2(9)) means a naturally occurring body of water that has been substantially changed from its natural condition as a result of *physical alterations* by human activity. More details of what this means are given in Annex II of the Directive. It could include a canalised section of a river, or a river that has been altered for flood prevention work so that the banks are straight and bare. It could include dredged water bodies where normal bed configurations are absent, or it could mean a stretch of river that has lost much of its natural flow due to the construction of reservoirs, or as a result of diversion into other water bodies. Member States must take decisions on when to apply this definition in determining water status.

Competent authorities

As with all other directives it is necessary to ensure that there is an administrative body set up to implement and enforce the provisions of a directive. In many such cases, because directives are addressed to Member States, the government often designates itself as the competent authority to ensure that this legal obligation is met. In other cases some individual organisation, not necessarily connected with government, may be designated, and the situation varies across the European nations. In the case of the EU Water Framework Directive, Article 3(2) and 3(3) states that the competent authority should be capable of applying the rules of the Directive to each river Basin District lying within the territory of the Member State and to any proportion of an international river which lies within the territory of the Member State. In the case of international rivers, an existing national or international body may be designated as the competent authority. There is therefore a need to ensure that the competent authority has sufficient standing to deal with all other organisations that have a bearing on the water in a river basin district, and in the case of international rivers, that trans-frontier negotiation and coordination is within the capability of the organisation. Countries will vary in these respects.

For example, in the mainland of the United Kingdom there are no international rivers as considered by the definition in the Directive, even though rivers cross the borders of England, Wales and Scotland, because the UK is a single entity as far as its membership of the EU is concerned. The UK can therefore choose to designate either one of its government departments (such as the Department of Environment Transport and Regions) or its two environmental regulatory authorities – the Environment Agency in England and Wales and Scottish Environmental Protection Agency as competent authorities in the knowledge that any possible minor disputes at the boundaries of the three countries are internal matters. In Northern Ireland the administrative system presently in place means that responsibility remains at central government level, and any discussions about rivers that cross into Eire are dealt with at government level. On the other hand, rivers such as the Rhine, pass through several independent Member States, including some where federal structures exist. In such cases, even within the countries there may be problems between regional governments. Some rivers such as the Danube pass through the territories of non-Member States, leading to more difficult decisions on choosing the appropriate organisations to be granted competent body status. The wide-ranging duties that are given to the river basin district organisations in respect of the implementation of other EU directives and the introduction of issues related to the recovery of costs within the context of the Directive add to the responsibilities of any such body. The ability to negotiate and cooperate with other similar competent authorities, even across national frontiers, is essential. Governments may wish to retain such duties within their foreign ministries, adding to the complexity.

River basin district

Essential to the decisions on where to place the responsibilities of the competent authority is the adoption of an appropriate definition for a river basin and its associated river basin district. The definition of river basin district is the *area of land and sea made up of one or more neighbouring river basins together with their associated groundwaters and coastal waters which is identified under Article 3(1) as the main unit for management of river basins,* so the districts become administrative areas designated for river basin management purposes. The area of the district may be larger or smaller that the river basins (or river catchment areas) that occur in its area. River basins are the areas of land through which surface waters flow to the sea and represent the physical attributes of the river basin district. Governments have to decide whether each river basin district should contain its own administrative unit to act as the competent authority or whether the districts should be managed from other central or regional structures. Further discussion of the river basin district concept is given in Chapter 5.

Water status

The Directive introduces the concept of "water status" and "good water status" qualified by additional identifiers such as *surface* water status, *groundwater* status and *ecological* status. This is a concept which is new to the world of water quality or quantity management and arises from consideration that natural environmental conditions vary throughout the EU area, and this must be taken into account in any water quality planning regime. The Commission has indicated in the preamble to the Directive that common definitions of water quality and quantity status should be developed, but that it should take account of local variations in geographical and other factors that affect the water bodies. Articles 2(17) to 2(28) cover a variety of definitions that are relevant to this concept, and the details of the parameters that must be taken into account are further explained in Annex V of the Directive. The *ecological* status of water includes reference to biological and morphological characteristics in addition to the more commonly used chemical classifications. *Chemical* status is defined in Annex V and includes reference to the achievement of quality standards in the water, particularly in respect of the absence of priority substances. *Good surface water status* is achieved when both the ecological status and the chemical status are at least "good". *High surface water status* is the condition of water bodies that are substantially unaffected by human intervention, and is the reference condition for specific types of water body.

The Commission and its experts have identified a number of water body types, characterised by common features such as altitude, size and underlying geology and geographical locations. The water bodies cover rivers lakes, transitional waters, coastal waters and groundwaters. Member states have to identify all their water bodies in each river basin district and place them into such categories. For each type of water body, a status representative of the conditions at the site in the absence of interference by man must be determined by reference to the ecology, chemistry and other parameters. Typical reference characteristics that represent high status, good status and moderate status are defined in Annex V. *Good ecological potential* is the term applied to artificial or heavily modified waters representing the possibility of attaining good status in due course. Groundwater is also assessed in terms of its *quantitative status.*

Priority substances

The Directive sets out to reduce or eliminate pollution by hazardous substances. The meaning of hazardous substance is given in Article 2(28a) but the Directive goes further to identify certain *priority* substances and *priority hazardous* substances that will be dealt with separately through Community level action according to Article 16 of the Directive; such substances are particularly damaging, and are listed in Annex X. Further details of the procedure to be undertaken are explained in Chapter 12.

Combined approach

Chapter 1 outlined the different methods that countries have adopted, to date, to control pollution. The Directive uses a combined approach and this is defined by Article 2(36) and in Article 10. The definition includes the establishment of emission controls that are based on the use of best available techniques for the industry concerned, or relevant emission limits set out in a number of earlier directives, and for diffuse sources of pollution, the introduction of systems of best environmental practices, together with the adoption of quality objectives where appropriate. The conditions which then apply to industry are the **more stringent** of the requirements for meeting the quality objectives or the emission limit values.

Water services and water use

Water services is defined as including all the activities that are relevant to the provision of water supplies and the disposal of waste water for households and industry. It includes abstraction, impoundment and storage, treatment and distribution of water and collection and treatment of wastewater. Article 1 indicates that sustainable water use is an aim of the Directive. *Water use* includes in its definition *water services* and any other activity that has a significant impact on the status of water, for example, abstraction, flow regulation, water transfers, artificial recharge, losses in distribution.

Emission limit values

Emission limit values that apply to discharges may be expressed as mass, concentration or levels of emission that must not be exceeded during any one or more periods of time. They apply at the point where emissions leave the installation. Where the discharges pass through a waste water treatment plant before entering the water environment, the limit values applicable at the installation may take the impact of waste water treatment into account. Thus emission limits could be less stringent, but this must not lead to higher levels of pollution in the environment, which must be protected as a whole.

Emission controls

There is a specific definition of emission controls in this Directive which may be different to that in others. The Directive points out that this should not be taken as a reinterpretation of the earlier definitions. The term means controls that entail a specific emission limitation such as an emission limit value or other conditions that affect emissions.

REFERENCES

Barcelona (1997)   O.J L 240 1977 p1
Helsinki (1994)   O J, L 73, 16.3.1994 p9
Paris (1998)   O J, L 104, 3.4.1998 p1

3

Principal obligations of the Directive

As is the case with all directives, the EU Water Framework Directive contains a number of primary obligations. These fall into a pattern similar to many of the existing environmental directives. There are obligations related to the initial transposition of the legislation itself; obligations concerned with initial data collection; the adoption of appropriate internal structures to carry out the work; establishing a suitable regulatory framework; monitoring obligations; and obligations to consult and report.

TRANSPOSITION

The terms of the Directive became European law on the date of its publication in the Official Journal, 22 December 2000 and Member States must transpose its provisions into their appropriate domestic legislation by 22 December 2003. The Commission must be informed when this is completed (Article 23). Transposition of the Directive may be quite complex as it is a framework involving links to a large number of other directives, not all of which are in the water sector. Some of the early water sector directives will be formally repealed over a period of years, as their provisions will be overtaken by, and incorporated within, the new Directive. In some cases existing laws may be adequate to deal with some of the requirements, as for example in France and the UK, where organisations operating within river basins already exist, with at least some of the powers to deal with issues on a river basin basis. Other countries have no such organisations, as pollution and water quality matters are dealt with in regional or local government organisations with administrative rather than catchment boundaries. The legal issues relating to the transition from old to new will be an interesting and potentially difficult area, requiring careful drafting of new domestic laws.

OBLIGATIONS RELATING TO DATA COLLECTION

Article 3 directs Member States to use river basins as the main planning tool for water management. It requires Member states to identify the water catchment areas from which water flows through streams, rivers and possibly lakes into the sea through a single river mouth, estuary or delta. Any associated groundwater and the coastal waters into which the river discharges are also added to the management regime in the river basin district, so these must be identified and delineated.

In addition to the gathering of information through conventional programmes designed to monitor quality and quantity, often known as surveillance monitoring, there are three new data gathering activities which must be undertaken in order to enable a river basin management plan to be formulated. Article 5 of the Directive requires an analysis of the characteristics of the river basin; a review of the impact of human activity on the status of surface and groundwaters; and an economic analysis of water use. Such surveys are essential to identify the main issues that affect water in the river basin in order to be able to consider measures for their control, and they must be completed within four years from the date of the adoption of the Directive. Article 5 applies to whole river basins and to those parts of international basins falling within the territory of the Member State if the river in question crosses international borders. There is a case here for international collaboration of a high order where important major rivers such as the Danube are concerned, in order to improve water status along the whole watercourse. The difficulties of applying common standards to the collection and interpretation of data by different countries may lead to considerable problems in the identification of problems and formulating their solutions.

ADMINISTRATIVE STRUCTURES

Once identified, river basins have to be assigned to a "river basin district". The river basin district is, under the terms of the Directive, an administrative region covering a geographical area defined as the river basin, and it is a requirement that the main organisation charged with water management issues for that area has the ability to take decisions relating to the area as a distinct unit. The river basin district, defined on a map, may comprise a single river catchment area, or it may encompass the water catchment or drainage area of several rivers. For administration purposes, the Directive allows the combinations of two or more river basins into a single district. Groundwater which is identified as lying under a river catchment area is assigned to the river basin district. Often, the boundaries of aquifers bear little relation to the surface topography of rivers, and are more closely aligned to the demarcations of geological strata. In such cases the groundwater is assigned, for administrative purposes, to the nearest or most appropriate river basin. Similar provisions apply to coastal waters which must be identified and assigned to river basin districts.

In the case of river basins that extend across international boundaries, the entire river basin must be administered as an entity. Such an obligation may require international collaboration, even with countries outside the EU. Existing treaties or collaborative arrangements may be brought into play to achieve this end, or where these are not in force, new arrangement must be made for managing the river basin.

Each river basin district must have appointed to it a competent authority to take the responsibility for implementing the provisions of the Directive. Within the territory of a Member State such administrative structures may utilise existing bodies, such as river boards, or environmental protection authorities, provided that each river basin district is managed as an entity within the organisation. Alternatively, entirely new structures may be established, the boundaries of whose areas match those of the designated river basin districts.

The situation is more complicated for those rivers that cross international boundaries. Here, provided the river as a whole lies within the EU it must be managed as a single entity (or the sub-districts that comprise the river basin). This will require coordination between the States involved. In many cases in the past, for example on the rivers Rhine, Meuse, and Elbe, this problem has had to be addressed as a result of pollution incidents, and international collaborative bodies have been established. The Directive permits the use of existing international bodies as the competent authorities in such situations. In some cases, European rivers are shared between Member States of the EU and states outside Europe – for example the Danube and its tributaries, such as the Maros/Mures and Morava. Article 3 requires each Member State to ensure that arrangements are in hand to apply the Directive to those parts of the river that lie within its boundaries, and to *endeavour* to apply its provisions to the remainder of the river through the establishment of coordination mechanisms with the non-Member State. There are examples of suitable organisations already in existence in Europe such as the Danube Programme Coordination Unit which acts between many Member and non-Member States, although the legal status of such voluntary collaborative entities will need to be examined further in the context of the Directive. Article 3(8) obliges the Member States to provide the Commission with a list of competent authorities by 22 June 2004.

PLANNING OBLIGATIONS

Because the Directive is forward looking towards improving the aquatic environment it contains a significant number of planning obligations. In order to identify the ultimate environmental state to which the provisions of the Directive are aimed, Article 4 specifies that objectives applicable to each water must to be elaborated by the Member State. The objectives have to define a condition for the water, which is known as "good status". This definition goes beyond any previous type of quality objective as it includes the consideration of factors other than simple quality parameters. A discussion of the issues involved in determining "good status" takes place in Chapter 7.

Waters used for the abstraction of drinking water are specifically identified, as one of the Directive's major aims is the protection of water resources. Where such waters are identified, environmental quality standards must be established. Such an action will permit the absorption of the current Directive concerned with surface waters abstracted for drinking water into the new Directive, but it will also include the need to give consideration to the quality standards of underground waters.

Other areas, which are designated as needing special protection, must also be identified through Article 6. This would include such areas as nature reserves where water plays an important part in maintaining the type of habitat and its related ecology. Such areas have to be listed in a register. The register of special areas must also include details of water bodies from which water is abstracted for drinking purposes and certain other areas identified specifically in Annex IV.

RIVER BASIN MANAGEMENT PLANS

A river basin management plan has to be drawn up for each river basin district. This will be based on actions needed to reach the environmental quality objective of "good water status", taking into account the results of water quality and quantity monitoring to establish the current situation, and the results of an in-depth examination of the impact of human activities within the river basin.

MONITORING OBLIGATIONS

The Directive specifies in some detail the type of monitoring that is required. Annex V, which was introduced at a late stage in the negotiations, sets out the monitoring programme essentials. Monitoring programmes must be established for surface and groundwaters and the protected areas.

REGULATORY OBLIGATIONS – PROGRAMME OF MEASURES

The river basin plan must include a "programme of measures" that takes account of the information obtained above. Article 11 of the Directive also stipulates that a number of basic measures must always be included in such a programme. These measures include:

- implementing the current EU legislation concerning the protection of water, using the combined approach which includes particularly controlling the impact of water pollutants by way of controls on their emission or the adoption and achievement of water quality objectives or standards. This can be achieved by the use of emission regulation through the adoption of best available techniques for the processes concerned or the use of fixed emission limits in other cases of point source discharges, and by using best environmental practices to reduce diffuse sources of pollution. These and other controls are set out in a number of environmental directives listed in Article 10 (for controls on emissions), and in Annex IX (for quality objectives) which must all be implemented within twelve years from the entry into force of the Water Framework Directive;
- cost recovery for water services;
- specific measures for the protection of waters used for producing drinking water;
- abstraction and impoundment controls;
- prior regulation schemes for point source discharges and diffuse sources based on prior authorisation, prohibition or registration, including the use of emission controls or general binding rules;
- a general prohibition of direct discharges into groundwater, subject to a number of exceptions;

- measures to eliminate pollution of surface waters by substances considered by the Commission to present an unacceptable risk to water, known as priority substances;
- measures to prevent or reduce the impact of accidental pollution;
- any other controls necessary to prevent deterioration of water. Annex VI contains a list of supplementary measures that might be used in this context.

There is an obligation for these measures to be put in place within 9 years from the commencement date of the Directive and made operational 12 years after that date, that is by 22 December 2012. Appropriate penalties have to be established to be used in cases of non-compliance with the national legislation which is invoked to support these measures.

OBLIGATIONS TO CONSULT AND REPORT

Consultation with interested parties including the public is an essential part of the new regime, and the competent authorities must allow the public to have access to the draft river basin plans, which should include a timetable and programme for the production of the plan, and information on consultation procedures at least three years before the plan period; an overview of the issues affecting the river basin two years before the plan period, and draft copies of the plan itself one year before it is activated, allowing also six months for comments to be made by the public. All background documents and information must be made available.

The Directive requires Member States to send copies of the River Basin Plans and any updates, to the Commission within 3 months of their publication, together with summaries of monitoring and analyses. Regular reports on progress will become an obligation.

The Directive is a framework and the implementation of all the other water directives become an obligation which is assumed to be achieved. In addition there are numerous other Community legal instruments that are linked to it, particularly in respect of their impact upon measures taken to protect water. These include the following:

- Integrated Pollution Prevention and Control Directive (96/61/EC)
- Habitats Directive (92/43/EC)
- Environmental Impact Assessment Directive (85/337/EEC)
- Seveso II Directive (96/82/EEC)
- Plant Protection Products Directive (91/414/EEC)

- Biocides Directive (98/8/EC)
- Landfill Directive (99/31/EC)
- Sewage Sludge Directive (86/278/EEC)
- Incineration Directives (89/429/EEC, 89/369/EEC, 94/67/EEC)

Although not likely to influence the programme of measures, the two directives concerned with public access to information must also be taken into consideration in the implementation of the EU Water Framework Directive: Directive on Access to Environmental Information (90/313/EEC), Reporting Directive (91/692/EEC) and Decision (94/741/EEC).

4

Implementation of the Directive

INITIAL ACTIONS

Transposition

In order to implement the Directive the legal requirement of transposition must
take place. The steps involved in this will vary from country to country,
depending upon what current legislation is in force to enable the various
obligations of the Directive to be introduced. It is possible that existing laws are
basically sufficient to allow the requirements of the articles to be met in the
practical sense, but it is most likely that countries will need to draft overriding
laws to bring their existing legislation into line with the Directive. Any laws or
regulatory instruments used to transpose the Directive must specifically mention
the Directive, and the measures have to be transmitted to the Commission. The
Commission usually scrutinises such laws to ensure that it is satisfied that
transposition is correct and complete. The use of administrative transposition,
used by several States in the past, is frowned upon. The Directive provisions
must form a part of statute law. Once transposition has taken place, practical
implementation follows.

Planning

The first key stage in the practical implementation of the Directive is the identification of river basin areas. Without completion of this task the provisions of the Directive cannot proceed. There is a dilemma facing governments on this issue from the start. Who is to identify river basin areas? Whilst it is clearly a government responsibility, it is a highly technical matter, for within river basins (watersheds or catchment areas), the Directive requires the inclusion of rivers and tributaries, lakes, canals, estuaries, certain coastal waters, and underground water. In many countries the responsibility for such diverse waters lies with different organisations, and the scientific expertise required to accurately identify and plot groundwaters, for example, may lie elsewhere. Should the government appoint a competent authority at this stage? In some countries of the EU this point is relatively straightforward to sort out, and river authorities, or environmental protection authorities, already exist in a suitable and unique form to be appointed. In others, where administrative boundaries form the basis of organisational structures, this is not so easy.

As a first stage, therefore, governments will have to decide upon the most appropriate organisation to delineate river basin boundaries. Until this step is complete it will not be possible, for technical reasons, to decide upon the appropriate size of the river basin district and the identification of the organisation which is to implement the provisions of the Directive. The Directive allows considerable discretion in sizing such Districts. The main criterion is that, once established, it must be sufficiently large to justify and support an organisation capable of carrying out, or at least organising and controlling, all of the functions appertaining to the management of water within the river basin area. This is a considerable task ranging from assessing the quality of surface and groundwaters to identifying and dealing with abstractions, and making economic judgements on various issues, sometimes at a cross-border level.

The government then has to establish river basin district administrations – institutions that are capable of carrying out the functions of water management in the identified river basin areas. According to Article 3, these institutions must be capable of carrying out the rules of the Directive, including international obligations where the rivers cross international boundaries.

Alongside the establishment of the administrative units, competent authorities have to be identified. This could be the individual river basin district administrations or it could be the government itself, or the responsibility could be granted to some other overseeing institution. The decision will depend to a large extent upon the existing water management structure and there may be a need to create new administrative structures, in order to achieve this requirement. Funds for the establishment or re-organisation of suitable organisations will be needed.

Once established and staffed, the first duties of the competent authorities will be to ascertain the current state of the river basin areas. Other duties follow from this. Article 5 requires that surveys be carried out on river basins and completed within 4 years of the coming into force of the Directive.

The surveys include an examination of the physical characteristics of the river basin area, with the intention of identifying what type of waters the basins contain (rivers, lakes and so forth) and what are the ecological characteristics of those waters, given the geographical and morphological positions of the river basins. This will allow a comparison of the actual ecological state of the waters with hypothetically unpolluted, and clean, waters of that type. A further discussion of what this means is set out in Chapter 6.

Surveys also must include an assessment of the impact of human activity on the basin. This will involve identifying point source and diffuse pollution, water abstraction sites where water is used for any purpose, the sites where flow regulation or diversion is in operation, and sites where changes in the natural morphology of the water body is taking place. Examination of other influencing factors such as land use patterns, including residential, industrial and agricultural use is required. Data applicable to surface and groundwater are included in the survey. Once the data have been accumulated, the impact of these activities must be assessed.

An economic analysis is required to assess the impact of cost recovery for water services in the river basin and to assess investment needs so that a judgement on appropriate changes to water use can be made.

As part of the initial survey, the river basin districts must identify any protected areas which lie within the river basin. These are areas specifically mentioned in other directives as requiring protection for nature preservation or other reasons and areas that are designated for the supply of drinking water. The latter are specifically mentioned in Article 7 of the EU Water Framework Directive.

Having gathered all the relevant information, the competent authority has to establish a set of environmental objectives for all the waters in the river basin. Further details of this requirement and its impact on water management are to be found in Chapter 7.

Diagrammatically, the first phase of implementation is shown in Figure 4.1.

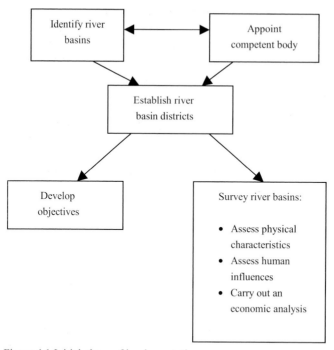

Figure 4.1 Initial phase of implementation.

The purpose of the surveys is to produce data from which the competent authorities or river basin districts can develop a river basin plan. However, the plan must itself lead to some tangible outcome, and that is set out in Article 4: environmental objectives.

As a result of the surveys, particularly using the results from the review of natural characteristics, Member States should be able to assign each body of surface water to an "eco-region" type. Why is this necessary? One of the main aims of the Directive is to bring all waters up to a uniformly good status in respect of their ecological quality. In other words, that all waters in the Community are of a condition that supports the appropriate flora and fauna and, in all other respects as to their natural characteristics, they are satisfactory. The problem with this aim is that the European Community area covers waters that reside in conditions from the near Arctic to Mediterranean in terms of climate, and a vast range of geological and morphological conditions. The natural, untarnished, status of a particular water body in the south of Italy will be very different to one in the north of Finland, and the ecologies of each water will not be comparable. Although the tendency of the European Commission is to look for standardised conditions for waters, it is not in practice possible to identify a single quality/quantity/ecology relationship that is common to all situations. The solution to this has been to develop a range of eco-region types differentiated by a number of common parameters, to which waters are ascribed, and leading to commonality of ecological qualities for waters which exist in similar situations. The detail of this will be explained further in Chapter 6, but for the purpose of the overall implementation plan, each body of water, identified as lake, river, sea, etc must be differentiated according to its ecological type – that is what ecology would be expected to occur given its geographical position on the European map, and its physical characteristics such as altitude, size, flow, geological basis and so forth. Such a "typing" will enable the competent authorities to determine what a natural quality should be, with no human interference.

SECOND PHASE

Objectives for surface waters

The Directive requires Member States to develop objectives for surface water bodies using the data derived above and the knowledge of the expected ecological quality, and then to prepare to meet these objectives over an agreed timescale by taking the appropriate environmental planning and protection measures. The overriding objective for all waters is the achievement of "Good status" within a timescale of 15 years of adoption of the Directive.

In terms of the overall implementation process, Member States have to decide upon what is "good status" for all of their water bodies. This is a new concept in respect of water management, for it is based mainly on biological population structures and compositions, and includes the requirement to ensure that all factors that affect this are taken into account. It will require a considerable investment in assessment techniques to determine these additional factors and to work out the relationships between them in order to come to a view of what is "good status".

Groundwaters

So far we have considered only surface waters. The Directive also covers groundwaters. The location and extent of groundwaters have to be identified. The biological condition for groundwaters is not such a relevant parameter in terms of its status. Instead, for groundwaters, the Directive concentrates on those parameters that indicate good quantity conditions – those which show that groundwater levels are stable despite abstractions; and those that indicate good quality status – such as the absence of saline intrusion, or that the concentration of pollutants is low, and that the groundwater quality would not cause a deterioration in any associated surface water. Having identified the quantitative conditions as a result of the survey the numerical objectives of "good status" for groundwater must be decided upon.

Artificial, modified and specially protected waters

There is some further planning work to be carried out before the final set of objectives is decided. Some waters have been so modified by human intervention that no amount of changes to inputs of pollutants, or alterations to physical conditions, will enable these waters to regain their original natural state. Furthermore there are many situations where bodies of water have been created artificially by man. In these situations it is often unrealistic to expect that new activities such as the removal of pollutant inputs, or changes to flow regimes will enable the water to be returned to its original state, or even to reach the "good status" definitions for comparable waters that are, at present, relatively unaffected by such interventions. In these cases a modified set of objectives may be applied, subject to overriding conditions (laid down in Article 3). These waters must therefore be separately identified.

At the other end of the scale, some waters need extra protection for reasons of the conservation of habitats or species, or because they are designated under other Community legislation (Article 6). Bodies of water that are used for drinking water supply or may be so used in the future are also picked out as being of special interest under Article 7. Both of these categories of water have specific requirements for their objectives.

Several factors may be taken into account that could lead to less stringent objectives being set. Article 4 permits the achievement of less stringent objectives, subject to certain conditions, if the achievement of good status would not be feasible, or actions needed to achieve "good status" would be disproportionately expensive. In these cases less stringent objectives may be proposed, but the reasons for this have to be explained to the Commission. Temporary deterioration from "good status" is not considered to be a breach of the Directive if this is due to natural causes such as prolonged droughts or floods, or other circumstances that could not reasonably have been foreseen and all practical steps are taken to mitigate the problems.

Programme of measures

Once all bodies of water have been examined and appropriate objectives set, the competent authorities have to devise a programme of measures to improve the situation and to meet them.

The programme of measures must include the adoption of all of the existing Community legislation that is aimed at the protection of water, together with a requirement to apportion costs to water users and to recover those costs, the introduction of measures which will achieve a sustainable water use in order to avoid compromising the attainment of the objectives, the adoption of measures to control the abstraction of water and the artificial recharge and augmentation of groundwater, and the introduction of systems for prior authorisation and control of activities connected with direct or diffuse sources of pollution. These measures are known as "basic measures". Member States can also introduce "supplementary measures" such as new legislative, administrative or fiscal instruments, codes of good practice, demand management measures and others to achieve the objectives.

River basin management plan

Article 13 requires Member States to produce a river basin management plan for each river basin district lying entirely within their territory. Where rivers cross international boundaries within the Community, a single international river basin management plan must be produced through coordination between states. Where this is not produced, each Member State must produce a plan for those parts of the river basin lying within their territory. If the river passes outside the Community the aim is to produce a single plan, but where this is not feasible, the EU State must at least produce a plan for the part of the river falling within its area. Supplementary plans may be prepared for parts of the river network, or to cover particular aspects.

River Basin Management Plans have to be prepared within 9 years of adoption of the Directive and must contain a description of the River Basin District including maps showing the bodies of surface water, the eco-regions and surface water types within the basins and their reference conditions for "good status", and the location and boundaries of groundwaters and protected areas. The objectives adopted for each water body must be included as must be the pressures and impact from human activity and a summary of the programme of measures to deal with these and hence achieve the objectives. Other items to be included are given in an extensive list in Annex VII of the Directive and discussed further in Chapter 9.

Each plan must be subject to public consultation before a final version is adopted, so the competent authority will have to make plans for suitable publicity and public access arrangements. The public must have access to information (not least as such issues are subject to the provision of the Aarhus Convention on public participation), and Article 14 of the Directive lays down specific consultation phases. Phase 1 is the release of a timetable and work programme including details of the consultation process 3 years before the plan period; a statement of the issues which will be dealt with by the plan, 2 years before the plan period, and finally, a draft of the plan itself must be released 1 year before the plan period. Six months must be allowed for written comments to be received before the plan is modified to its final state.

The outline for this second stage of implementation planning is shown diagrammatically in Figure 4.2.

The competent authorities must provide for reviews within this overall plan. The survey of the river basin must be initially completed within 4 years of the coming into force of the Directive, and reviewed for the first time within 13 years with a further review every 6 years thereafter. This will enable the river basin district the opportunity to see the effects of their activities. The river basin plan itself is to be published within 9 years of the adoption of the Directive, and reviewed and updated at the latest 15 years thereafter. The plan must then be reviewed every 6 years.

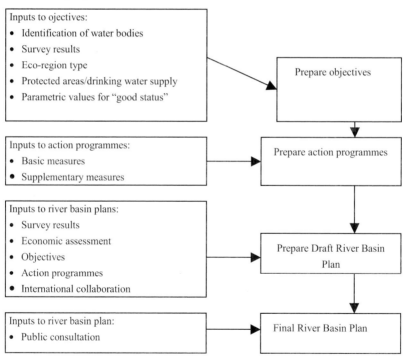

Figure 4.2 Second phase of implementation.

The third stage of implementation might be considered to be the "operation and maintenance" stage. The competent authorities must activate the plan and be in a position to assess the state of the waters under its control, in order to judge the effectiveness of the measures they are taking. They must also have sufficient powers available to ensure that persons required to take action do so. Thus the important continuing duties concern the monitoring of water status and enforcement of the legislation.

THIRD PHASE OF IMPLEMENTATION

Activation of the plan

Using the legislation brought in to transpose the Directive and all other available legislation the competent authority must activate the practical measures outlined in the plan in order to achieve good water status within the timescale. This will involve fully implementing the programme of measures. The basic measures will be achieved by ensuring that all existing directives are fully implemented, that there is suitable funding and staffing to meet their requirements. The permitting systems must be introduced and measures taken to ensure that all the necessary controls on dischargers and abstractors are applied to reduce the impact of human activities on water bodies.

Enforcement

The achievement of "good status" is likely to be achieved only if all of the measures identified in the river basin plan are carried out. Measures such as regulatory control of discharges, or prohibition of certain activities may be unpopular. Compliance with emission limits or other conditional measures may be difficult to achieve, and it is possible that they may be breached for a variety of reasons. In order to encourage compliance Article 23 of the Directive requires Member States to adopt penalties for breaches of the measures put in place to implement the Directive. These may consist of the current regulatory system for water regulation, and industrial pollution control, and all other activities that affect water quality and quantity, but it must made be clear as to how they relate to the Directive.

Monitoring

The competent authority must establish a monitoring programme to determine water status. Monitoring has to include the assessment of chemical and biological quality, and must be capable of differentiating the three classes into which surface waters may be placed as a result of the surveys described earlier - that is "high status", "good status" and "moderate status" - and two further classes "poor status" and "bad status" which may be identified during the monitoring programme. Each of these classes is used to describe the ecology, the hydromorphology, the physico-chemical quality and biological quality of the identified bodies of surface water in each river basin, that is, rivers, lakes, transitional waters and coastal waters. It is clear that the monitoring programme is fairly complex and will require the establishment of comprehensive sampling and analyses facilities in a number of scientific fields. For groundwater the monitoring programme consists of quality and quantity parameters. Quantity is related to a groundwater level regime requiring the setting up of a groundwater level monitoring network whilst quality is ascertained by means of a programme to measure the chemical condition of groundwater. Annex V sets out the detailed requirements for monitoring programmes and these will be discussed further in Chapter 13. Monitoring frequencies vary from one sample in 3 months to one sample per 3 years.

A long-term requirement, stemming from the aim of the Directive to produce a common set of standards for waters of each specific eco-types so that common definitions of "good status" may be derived, is the need for Member States to arrange to exchange data in order to build up a picture of each eco-type.

Recording and reporting

A recording and reporting system is needed to collect and store all of the data collected from the river basin surveys, the consultation processes and the monitoring of water status. Many countries already have such systems, but often these are organised on a discrete sectoral basis. This Directive is integrated in its approach and in its aims, and requires extensive public access to data. There is a heavy reporting requirement both to the public and to the European Environment Agency that must also be accommodated. Member States will have to establish a suitable recording and reporting system for this purpose. Where international river basins are established, the recording system will need to be compatible between Member States, and if the river basins extend beyond the boundaries of the EU, consideration of the type of recording systems will have to take this into account.

Reports must be prepared for the Commission on river basin districts, which include the groundwaters and coastal waters assigned to each district; competent authorities; issues that affect water management but which fall outside the area of competent of those authorities; river basin plans for complete river basins and sub-basins or other localised plans; and also for international river basins.

The third phase might be described by Figure 4.3.

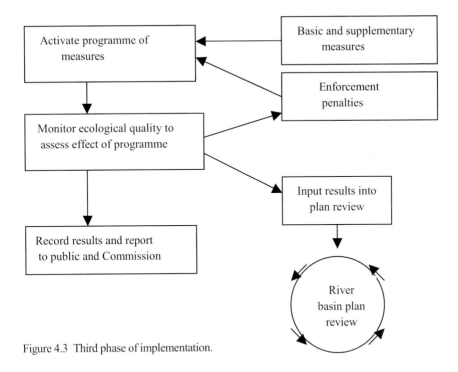

Figure 4.3 Third phase of implementation.

OTHER ISSUES

There are a number of other issues identified in the Directive that will be addressed concurrently with the above main phases. Particularly important issues relate to the need for coordination between Member States in relation to river basins which cross international borders, and some specific tasks assigned to the Commission for development at Community level.

Community level tasks

These include:
- further definition of eco-regions and types of water bodies together with suitable reference conditions, and criteria for designating heavily modified bodies of water and the definition of their maximum ecological potential;
- identification by the Commission of priority substances and derivation of environmental quality standards for surface waters, with the development of Community-wide measures for emission control of these substances;

- development of EU criteria for identifying what may be considered to be significant anthropogenic impact;
- formulation of possible means for controlling diffuse pollution on a Community-wide basis;
- development of methods for economic analysis and long-term forecasting.

International river basin districts

Where the river basins cross international frontiers between Member States and also other countries outside the EU, such as the River Danube, special attention will have to be given to some additional implementation tasks. Within the Directive these tasks are identified in each of the appropriate articles, and it will be necessary to include the extra work in the three phases described above, alongside the internal activities. The main additional factors include:

- identification of international river catchment boundaries and areas, and assignment of those parts of the river basins that fall within individual Member States' territories to international river basin districts;
- establishment of suitable international administrative organisations between the Member States involved;
- identification of those protected areas that extend into other States territories according to Annex IV;
- agreeing common environmental objectives between States for the whole river basin (Article 4);
- identifying and mapping the ecological quality of surface waters in collaboration between States;
- collaborating on monitoring programmes and sampling sites,
- coordinating the review of river basin characteristics
- coordinating the programme of measures and identification of pollutants
- coordination of the preparation of an international river basin plan and making arrangements for consultation beyond the confines of the individual Member State's territory to cover the whole river basin;
- development of suitable joint reporting procedures.

5

River basin districts

ARTICLE 3

Article 3.1 of the Directive requires Member States to *identify individual river basins within their national territory* and to assign them to *individual river basin districts*. The fundamental issue here is that the management of water as a whole is to be based on river basins. The river basin district may comprise the watershed of a single river, or an area that contains the catchment areas of several smaller rivers. Most countries are in possession of information regarding the catchment areas of their rivers and, given the known influence that many of the geographical features of river basins, and human activities within them, have on the quantity and quality of water in the rivers, management decisions are often based on the requirements of the basins. The Directive extends this type of approach to all rivers, and includes underground waters which occur within the river basins, the estuarial waters at the river mouths, and an area of sea that could be affected directly by the river discharges.

Member States have to set up administrative arrangements under which a competent authority is able to apply the rules of the Directive on a specific river

© 2001 IWA Publishing. *The EU Water Framework Directive: An Introduction* by Peter A. Chave.
ISBN: 1 900222 12 4

basin district basis. The Directive does not necessarily require Member States to establish management organisations within each of its river basin districts. It does not specify any particular management structure, although in several countries such institutional arrangements already exist.

Where the river basins extend over international boundaries the river basin districts must be extended to take this principle into account, and suitable arrangements sought between States inside and outside the Community to permit the management of these rivers on a "whole" river basin basis.

IMPACT OF THE DIRECTIVE

There are two distinct, and major, implementation activities introduced by this Article. The first is the identification, using geographical and hydromorphological information, of the catchment areas or watersheds of all rivers in a country's territory, a mainly technical activity. The second is the establishment of the necessary management arrangements for the river basins. The identification of an appropriate organisational structure to manage each basin, or the establishment of new relevant institutions, will depend upon what is already in place, but it is necessary to decide whether individual river basins are sufficiently large to warrant the establishment of appropriate organisational structures for their management or whether basins should be joined together for administrative purposes. No guidance is given on this aspect in the Directive, but as the Directive is addressed to Member States it becomes a government responsibility to ensure that the right decisions are taken at this stage. The timescale is short as Article 3(7) requires that the competent authorities for managing river basins must be identified by December 2003, and these cannot easily be established until the river basin districts have been decided upon.

IDENTIFICATION OF RIVER BASINS

Rivers are commonly managed at a river basin level. It is generally thought that activities within a river basin tend to affect the river that drains that basin, and that if the quality or quantity of water in the river is to be subject to a managed regime, then those activities within the river basin should be the ones subject to management control. Activities that lie outside the catchment area tend not to be as important to the specific river. This is not completely true, as, for example, airborne contaminants can affect rivers over large distances, and sometimes activities in one catchment can lead to effects in adjacent or more distant catchments, for example, the movement of solid wastes from their area of production into an area of disposal in another river basin, or even the transfer of water from one river to another to supplement supplies. Furthermore, groundwaters do not always follow precisely the demarcation of the river basin although they may have an important effect on base flows and possibly the intrinsic quality of river water. Nevertheless from the point of view of establishing a management regime, the river basin is of greatest importance.

Under the terms of the Directive, whilst no guidance is given as to the appropriate size of a river basin district, there is an assumption that the district will be sufficiently large to enable effective controls to be introduced and that the administrative organisation will be sufficient to enable the rules of the Directive to be properly applied. Any organisation set up to act as the administrator of the river basin district will have a large number of tasks to perform and it is essential that it is large enough to accomplish these. This means that river basin districts will need to be of a sufficient size to sustain staff and expertise, and to be able to raise sufficient funds to carry out this work. The Directive allows Member States to combine a number of small river catchment areas into a larger district for such purposes.

The reality of the situation in Europe is that the watersheds of most rivers are already recognised and defined on a geographical basis, and this information is readily available on maps or in other forms of geographical information system. However, the Directive aims to place control of all matters affecting individual river basins within the watersheds of the rivers. Assurance will be needed that the precise demarcation between river drainage areas is known. Although the channels of rivers are known, in some countries further surveys may be required to ensure that accurate watershed extremities are identified.

There will also be additional tasks to determine the extent of ground water and to decide into which individual river basin such aquifers should be placed for administrative purposes, for it is not always the case that underground water flows follow surface watercourses and river basin boundaries and in many countries groundwater is currently not managed on a basis that is related to rivers, so that links will need to be established.

Similarly, coastal waters which fall within the definitions in the Directive will have to be identified and assigned to an appropriate river basin. Where there are several rivers discharging to the sea along a coastal stretch, decisions will be required as to which area of sea is most influenced by which river. Such decisions become important where future improvement works are necessary to deal with particular water quality issues.

In situations where no formal organisation is already in place to undertake the task of assimilating available information and collecting the new information on ground water and coastal waters, this task will fall to central government or to a body given the task.

RIVER BASIN DISTRICTS

Once individual river basins are identified, along with their associated aquifers and transitional and coastal waters, they must be assigned to a river basin district. The river basin district will become the area within which administration of the individual river basins will take place. The areas of river basins range from the very large such as the River Danube with an area covering 817,000 square km and absorbing whole countries (such as Hungary), to the very small (such as rivers in Cornwall, UK) where the entire river comprises merely a few kilometres from source to sea. The Directive takes account of such huge variations by enabling river basin districts to be identified that comprise a number of small river basins, or where the geographical watersheds of small rivers are joined to those of large rivers for administration. There is no facility to subdivide the administration of very large rivers (into tributaries, for example). However, at a practical level, sub-basins may be the subject of separate plans (see Chapter 9).

Because the river basin district will become the focus for action to implement the main provisions of the Directive, it must be of a sufficient size to allow of funding and staffing to accomplish this – it must be of a strategic size. There are additional problems associated with the river basin districts of international rivers.

Having identified the river basin districts the governments of Member States must *"ensure the appropriate administrative arrangements, including the identification of the appropriate competent authority, for the application of the rules of this Directive within each river basin district lying within their territory"*. What does this mean in practice?

Models for river basin districts

The definition of river basin districts in Article 2(15) is that it is the area of land and sea identified as the *"main unit for management of river basins"*. For each of these units there must be appointed a competent authority. There are several models for Member Sates to consider in the application of this requirement. These will depend upon the present administrative structure and culture of the particular country. Some countries already have experience of operating a water management regime based upon river basins and may chose to simply use their existing administrative structures. For other countries there may be a need for institutional reform. France and the UK are two countries in which current water management regimes may largely fit the new situation, but there are other possible models that could be adopted.

Agences de l'eau

In France the institutional arrangements for dealing with water cover a variety of organisations including the central state and the municipalities. The central state has general responsibility for water resources, the authorisation of abstractions and discharges and drinking water quality, and operates through a variety of separate ministries (Barraque *et al.* 1998). A regional tier of government has a role in implementing the policies of the Ministry of Environment including monitoring of water quality, and is also involved in water planning where these cover areas that are larger than the geographic limits of a *départment.* The *départment* is the key institutional level at which overall water policy is implemented through the coordinating activities of the Prefect. However there is a particular feature of the French institutional system that fits well with the new concept of river basin districts. Following the adoption of a 1964 law, *Agences de bassin* were established in 1970 with the aim of preparing plans and incentive policies (based on financial initiatives) for river basin management. The six *Agences* were reorganised under a 1992 law into *Agences de l'eau* and they cover the entire French territory in a way that reflects the six main river basins in France – the Seine, Loire, Rhône, Adour, Artois and a basin based on the international rivers Rhine and Meuse. The *Agences* have responsibility for the rational use of water in the basins but they do not play any part in discharge or abstraction controls nor in infrastructure construction (although they can finance this through contracts). In the terms of the new Directive, the French situation has the immediate advantage as the definition of river basin areas is already well established organisations are available that have knowledge of the basins. Also, to a large degree, the size with which to achieve river basin planning, the experience of the *Agences de l'eau* should be a useful starting point for implementing the requirements of the Directive. However, there is much more to the Directive than simply drawing up plans and financing certain operations, and a significant move of, for example, the control of installations under the *Classified Installations* Regulations (CI 1997), which is currently the responsibility of municipalities (through the *Prefect*), may have to be considered. At the least, links between the water planners and the pollution regulators would require strengthening. Figure 5.1 shows the approximate extent of the areas of the *Agences.*

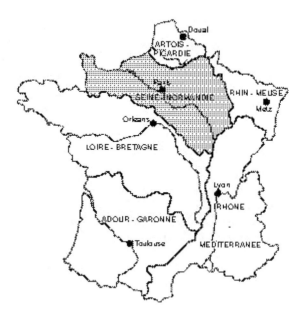

Figure 5.1 *Agences de l'eau* in France.

The UK Environment Agency

In 1973 an Act of Parliament (UK 1973) was passed which introduced a new concept in water management in England and Wales. As a result multifunctional *Regional Water Authorities* (RWAs) were established in 1974 from an amalgamation of local water boards, sewerage undertakers, and river boards. The new bodies had boundaries which were carefully drawn to match the watersheds of large rivers such as the Thames, the Severn and Trent, or where no single large river existed, a number of smaller rivers were conjoined into a single area. Ten such water authorities covered England and Wales. The new concept meant that water policy could be developed and implemented on a catchment basis. The new bodies were given wide responsibilities for water management and this included all aspect of natural water from floods to droughts, from fish protection to recreational use. The inclusion of the organisations that abstracted and supplied drinking water (water boards) and the organisations that collected and treated waste water and then discharged the effluent to the river network meant that there was control over the entire water cycle – an innovative management regime. The advantages of such an institutional development were seen as providing a means of integrating the management of water resources and water quality over entire catchments, including the ability to take proper account of the priorities for expenditure; and the establishment of bodies sufficiently large to be able to provide the right level of expertise, laboratories, design teams and so forth. Such water authorities operated for some 15 years. A major problem during this period, however, was the inability to obtain sufficient capital for large construction projects at a time when new EU legislation required massive investment in infrastructure. As the authorities were public bodies that required funding from central government, there was political concern at allowing charges for water and sewage treatment to rise to the levels needed to cope with both public expectation of improved services and to meet the new legislative commitments.

In 1989 the government decided to privatise the supply and treatment operations as a means of introducing private finance into the infrastructure development, and it disassembled the RWAs in favour of a new institutional framework comprising private companies who would supply water and sewerage services and a new regulatory organisation for protection of the water environment – the *National Rivers Authority* (NRA). The new body was separated from government as a non-departmental public body with its own board of management, and it was responsible for all the regulatory and flood defence functions of the previous RWAs. Although the new body was a national organisation, it retained a regional management structure based as before on the ten major river catchment boundaries. Thus catchment management planning could be introduced without difficulty by each of the regional offices. For a perceived administrative convenience at the time, the geographical boundaries of the new private water companies were also based initially on approximately the same borders. Thus water regulation and the provision of water based services were still linked geographically and investment plans could be effectively catchment based. Figure 5.2 shows the catchment boundaries of the NRA.

Figure 5.2 Boundaries of the NRA.

A further reorganisation of the regulatory body has recently taken place in the UK, partly as a result of recognition of the growing importance of taking an integrated approach to environmental management. The Environment Act of 1995 (UK 1995) has established an independent environment agency in England and Wales and a similar body for Scotland (known as the Environment Agency in the former, and Scottish Environmental Protection Agency in the latter). Both agencies have absorbed the former water regulatory and management bodies, and have added to these the organisations that regulate industry under integrated pollution control, and the solid waste regulatory bodies. The duties of the new bodies are much more extensive that the former individual organisations. The Environment Act defines the Agency's aims as *"so to protect or enhance the environment, taken as a whole, as to make the contribution towards attaining the objective of achieving sustainable development (that) Ministers consider appropriate"*. In that its principal aims refer to sustainability and the integrated approach, these match closely the aims of the new Directive as set out in the preamble.

The new organisation, although subject to central control, is organised on a regional basis for the purposes of environmental management. The regional boundaries are an amalgam of the boundaries of the former National Rivers Authority (which were based on river basins) and the former industrial pollution regulator (Her Majesty's Inspectorate of Pollution (HMIP)) whose boundaries were administratively determined. The new regional structure is based upon eight rather than ten regions, but their boundaries remain largely based upon river catchment boundaries. The view was taken at the reorganisation that many aspects of pollution and environmental management influenced the environment through drainage basins related to rivers. The structure thus has the potential to deliver the requirements of the EU Water Framework Directive with little revision.

Italy

In Italy a reorganisation has recently taken place to establish a rather similar administrative system for river management. A National Environmental Protection Agency was established in 1994 with extensive responsibilities for environmental supervision and, more recently, river basin authorities have been set up covering 6 basins of national importance as well as 15 regional and 17 inter-regional basins.

Federal states

Germany is an example of a federal republic. Legislative competence is held by the individual states – the *Land* – in addition to the federal government. In terms of determining the identity of the competent body for this Directive the Directive is addressed to the federal government and it will have to make the decisions on the appointment of competent bodies. The *Land* governments are accountable to *Land* parliaments and have authority for activities within their *Land*. At the central federal level a variety of ministries are involved in water issues. The Federal Ministry of the Environment, Nature Conservation and Nuclear Safety is the lead ministry, responsible for water resource management. It controls a number of federal agencies including the Federal Environment Agency. However, certain other ministries have specific responsibilities in the water field. Rural area water resource issues are dealt with by the Ministry of Food, Agriculture and Forestry whilst the Ministry of Health deals with drinking water quality.

As a federal republic the central, federal, ministries have to work through the governments of the individual states. It is here that water planning is mainly undertaken. Each of the *Lander* has an environment ministry or similar which has the responsibility of a supreme water authority although there are detailed differences between the *Lander*. Regional governments serve as higher water authorities whilst rural and city districts also play a role in water management as lower water authorities (Kraemer and Jager 1998). A generalised diagram is given in figure 5.3.

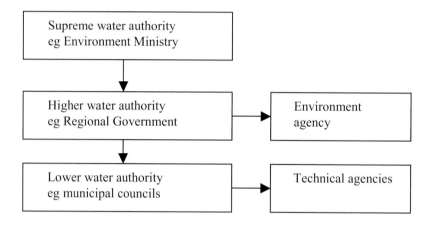

Figure 5.3 *Land* Authorities in the German water management regime.

In some areas technically competent environment agencies are appointed to carry out the work on behalf of the regional or local governments, and these are supervised directly by the *Land* Environment Ministry. As is evident, the organisational structures in these states are not based primarily upon river basins but on local administrative boundaries, although the *Lander* have established cooperation procedures between them to encourage river basin management, and also the establishment of Commissions to deal with problems on rivers which cross international boundaries. From the point of view of this Directive, the institutional structures in Germany will require significant alteration in order to operate on a river basin basis.

United States

The position in Germany is not dissimilar to experience in the United States. Here there is a federal structure with many organisations (for example, the Environmental Protection Agency) which operate with a nationwide remit under federal laws, but where individual states retain responsibility for many of the practical management issues connected with their internal rivers. In many cases the large rivers of the US pass through several states and arrangements must be made for joint management regimes to be set up. The Delaware River is one example of a collaborative approach between individual States which all have their own individual laws and regulations yet have agreed a management structure to cover the whole catchment (US 2001). The Delaware River Basin Commission was formed in 1961 as an equal partnership organisation between federal government and a group of States for the purposes of river basin planning, development and as a regulatory agency. The river is fed by 216 tributaries, and its 13,539 square miles cover the states of Pennsylvania, New Jersey, New York and Delaware. Commission members are the governors of the four basin states and a nomination from the President. Before its establishment some 43 State agencies, 14 interstate agencies and 19 federal bodies dealt with various aspects of river management. The responsibilities of the Commission are laid down in a statute, so that its actions have a legal basis. They cover water pollution abatement and regulation of discharges, water supply allocation, water conservation, regional planning, drought and flood management and recreation. The Commission has prepared a series of comprehensive plans since 1962. The plans have allowed the Commission to clean up the river as a whole to a high quality standard and this has benefited fishing and recreation. It has set its own regulations covering such issues as water conservation, restriction of development in flood plains, metering of large water abstractions, protection of high quality waters in parts of the watershed, the discharge of toxic pollutants from waste water plants, and protection measures for groundwaters. The Commission is an example that government bodies, each with their own sovereign powers, can work together on an equal footing given the appropriate collaborative approach and the right legal basis. The Delaware basin is depicted in Figure 5.4.

Figure 5.4 Delaware river basin. (Reproduced with kind permission from the Delaware River Basin Commission).

Portugal

In Portugal, river basin management is organised on a national and regional level. A national ministry – the Ministry of Environment and Natural Resources – has overall responsibility for water policy. It is supported by a number of coordinating and sectoral organisations. In the water field the key organisation is the National Institute for Water (INAG) which has the responsibility for water planning and coordination at national level. In the regions are five separate Regional Directorates for the Environment and Natural Resources (DRARN). These bodies are responsible for all activities relating to water, air, waste and nature conservation. Eleven of 15 river basin plans are the responsibility of the DRARNs whilst INAG must produce the national plan and river basin plans for the four rivers that require significant cross-border coordination. In terms of the new Directive the content of plans will conform readily with the requirements, but there are a number of options over the appointment of the competent authorities. The five DRARNs are administrative districts whose boundaries do not match those of river watersheds (Correia *et al*. 1998). There will thus be problems in determining which organisation should be the competent authority for each river basin. Should the INAG be made the competent body or should there be a reorganisation to match the river basin boundaries? A further interesting development is the appointment of 15 River Basins Councils and a National Water Council to supervise the preparation of plans. What role should these bodies play in competency?

Figure 5.5 River basins and administrative structures in Portugal.

INTERNATIONAL RIVER BASINS

A particular problem which will affect most but not all of the EU Member States is the difficulty of management of river basins that cross international frontiers. The Directive, at Article 3(3), requires Member States to *ensure that a river basin covering the territory of more than one Member state is assigned to an international river basin district.* There are several examples of international rivers that have for many years been subject to cooperative measures between riparian states. However, at the level of intervention suggested by the new Directive particularly in respect of the programme of measures, significant work will be required to improve collaboration. Some examples of such cooperative ventures include the well known international commissions for the Rivers Rhine and Danube.

River Rhine

The basis for international cooperation on issues connected with the Rhine is the 1963 Agreement on the International Commission for the Protection of the Rhine Against Pollution. The River Rhine has a catchment area of about 185,000 km^2 spanning ten countries, and its tributaries are major rivers in their own right. That part of the catchment area lying in France covers two regions: Alsace, which drains directly, and Lorraine, which drains via the Saar and Moselle which enter the Rhine in the Federal Republic of Germany. Part of the Rhine forms the national frontier in Alsace and in administrative terms the *Agence de l'eau* of Rhine–Meuse is also involved, as this river basin authority deals with rivers which drain to the North Sea, including the Meuse and those stretches of the Saar and Moselle and Rhine which lie in French territory. Germany, France and Luxembourg have established separate protocols to cover the main tributaries, the Moselle and the Saar, as the river basins are shared between these countries. Several contractual agreements have been signed between two German *Lander* and Alsace concerning groundwater and flood protection. In the context of the new Directive, such agreements may form the basis of the competent body for applying the rules of the Directive. However, the complexity of assigning responsibility for all of the provisions becomes clear when it is considered that at present these are voluntary agreements made between independent states, and they cover only very specific issues such as monitoring programmes, sharing of data, and agreements to control particular pollutants.

River Danube

The Danube presents an even greater problem for determining the river basin district and the appropriate competent authority. The Danube is the largest river in Europe covering a catchment area of 817,000 km^2 with a main river length of 2,857 km. The river and its tributaries pass through 17 countries in total. Its watershed includes the whole territory of Hungary, much of Austria, Croatia, Romania, Slovakia, and Slovenia, and large parts of Bulgaria, the Czech Republic, Germany, Moldova and the Ukraine, and small parts of a further seven countries are also included within its basin. The principle of river basin management in such a massive river and the institutional complexities are clearly enormous. Nevertheless, the river holds such significance for the environmental, economic and cultural well-being of these countries that already the countries concerned have moved towards basin-wide cooperation.

The Danube is a heavily used watercourse for commercial navigation, drinking water supply, agriculture, industry, fishing, recreation and hydroelectric power generation. Individual riparian states have developed many dams, locks and other hydraulic structures for their own benefit. Although the developed western countries had made progress in reducing pollution and environmental damage, those in central and eastern Europe did not develop or implement such policies. As a result the waters of the river and its tributaries are degraded in many ways. Recognising the need for action, in 1991 an environmental programme for the Danube was drawn up by a meeting of interested countries and other parties in Sofia. In order to implement this programme, the countries of the Danube river basin and the EU signed a Convention on Cooperation for the Protection and Sustainable Use of the River Danube, in 1994. The Convention is aimed at achieving a sustainable and equitable management of the river, and the signatories have agreed to cooperate by taking *all appropriate legal, administrative, and technical measures to at least maintain and improve the current water quality conditions of the Danube River and of the waters in its catchment area and to prevent and reduce as far as possible adverse impacts and changes occurring or likely to be caused.* The Convention also sets up an International Commission for the practical enactment of the Danube Environmental Programme.

In the context of the EU Water Framework Directive, this Convention must be regarded as the precursor to the identification of an appropriate competent authority. Article 3(3) indicates that the European Commission is available to act as a facilitator in the decisions regarding the assignment of the river basin to an appropriate river basin district. However, it is not clear whether a commission set up under such a convention could legally act as a competent authority as it is not subject to the law of an individual Member State. Furthermore, in the case of the Danube, the Commission includes signatory countries outside the EU. These are not subject to EU legislation. Indeed of the 18 countries in the Danube basin, only three are current Member States and a further five are accession candidates at the present time.

The difficulties of identifying an organisation with true responsibility for implementing the Directive requirements are immense in such a case. Whilst the provisions of Article 3 (5) – the establishment of coordination with non-Member States, and the application of the rules within the territory of the

Member States concerned – is feasible and desirable, the overall effectiveness of applying the river basin concept will rely heavily on goodwill.

International coastal waters

Although coastal waters encompass only the strip of water one nautical mile wide from the shoreline (Article 2 (7)) and lies fully within territorial waters,the need for international cooperation becomes apparent in the cases where this water may be affected by a river which forms an international land boundary and the coastal water is thus divided between the riparian states. It is also necessary where the coastal strip is adjacent to a territorial boundary. Examples of these might be Minho and Guadiana (Spain/Portugal), the Rhine Holland/Germay and the Ems (Germany /Poland). The Danube presents a special case, for its transitional and coastal waters fall in countries outside the EU, yet under the Directive must be included in the river basin plan. In some cases it may be possible to utilise the framework of existing treaties as the basis for dealing with such water bodies.

IMPLEMENTATION ACTIVITIES

Member States have a number of urgent tasks in respect of this part of the Directive. They must decide how to identify the river basins, including international river basins, carry out the necessary work to demarcate the catchment boundaries, examine the sizes of the river basins, and decide whether they are appropriate for setting up river basin districts. At this point alternative sizes may be proposed, so alternative river basin management organisations and options for competent authorities need to be assessed, including the viability of the options prior to decisions being taken. All this must be completed in time for the deadline of 22 December 2003.

Several questions need to be answered. Who will carry out the initial surveys and assessments? Who will undertake surveys of underground waters and coastal waters to determine their correct placement? There will be a need for contact to be made with both upstream and downstream countries. Who will arrange for international collaboration at this early stage? It could be that existing organisations could cope, or entirely new ones established. The problem is likely to be that existing structures already have a full range of duties to perform, and the Member States' governments will find it necessary to take this into account when beginning the task of implementation.

REFERENCES

Barraque B., Berland, J-M., Cambon, S. (1998). In Correia, F.N (ed.) *Institutions for Water Resources Management in Europe*, Balkema, Rotterdam.
Correia J.N, Neves, B.E, Alzia M, Da Silva S&J.E (1998), idem
Kraemer R.A, Jager F (1998), idem
CI (1997) Classified Installations, Ministry of Environment, Paris.
UK (1973) The Water Act 1973 (c37) HMSO, London.
UK (1995) The Environment Act 1995(c25) HMSO, London.
US (2001)

6

River basin characteristics

OBLIGATIONS IN THE DIRECTIVE

The first task of a river basin district authority (or other competent authority) will be to collect information about the river basin. Within four years of the adoption of the Directive – that is by December 2004 – Member States must ensure that the provisions of Article 5 are completed. Article 5 requires all the details of the natural characteristics of each river basin to be recorded; the impact of human activity of the status of surface and ground waters in the river basin must be assessed; and an economic analysis of water use in the river basin must be completed. This information is a vital part of the river basin planning process and the Directive specifies that it must be updated within 13 years of the date at which the Directive comes into force, and at intervals of six years thereafter.

© 2001 IWA Publishing. *The EU Water Framework Directive: An Introduction* by Peter A. Chave. ISBN: 1 900222 12 4

REVIEW OF NATURAL CHARACTERISTICS

Annex II prescribes in some detail how to go about characterising the surface and ground water bodies in each river basin. Surface waters must first be identified, and classified as rivers, lakes, transitional (estuarine) or coastal waters, or as artificial water bodies (for example reservoirs, or canals) or in some cases as natural water bodies that have been heavily modified in some way by human intervention so that their natural characteristics have been altered. Precise definitions of the different categories of surface water body are set out in Article 2 of the Directive (Definitions). Groundwaters are identified according to their location, their boundaries, and uses, together with an assessment of whether they may fail to meet their objectives (described in Chapter 7). Although countries will be aware of the location of most water bodies, the Directive requires more detailed information than is often available (for example the precise boundaries of presently unused groundwater bodies) and, more importantly, a database of such waters will need to be established. This is because, in due course, a map in Geographical Information System (GIS) format showing all the water bodies has to be submitted to the Commission.

Identification of water bodies

Rivers

The first task is to identify the water bodies. River basins contain innumerable water bodies, ranging from main river channels, through tributaries to ditches. At what point should there be a cut off in the map of river channels? Arguments over this raged for many years in the UK when its Rivers (Prevention of Pollution) Act of 1951 and several subsequent acts attempted to define "controlled waters" – those waters which were to be regarded as coming under legal protection for the purposes of controlling effluent discharges into them. The outcome of this argument led to the decision to treat any surface watercourse or underground water which received discharges as a controlled water. Annex II of the Directive appears to limit the size of rivers which might be considered as water bodies to those with a catchment area of greater than 10 km^2. It will be a considerable exercise to identify such waters. A further difficulty is the decision on what length of a river or stream constitutes a water body for the purposes of determining its relevant pristine ecology. A typical river will have a wide range of hydromophological and physico-chemical conditions along its length by virtue of the fact that rivers usually (but not exclusively) rise in upland situations and pass to the sea through a variety of geographical and geological situations, all of which will affect the type of ecology which should be present. Annex II also seems to indicate that rivers with a catchment area of greater than

10,000 km^2 would still be categorised as a single water body. This seems to contrast with most existing systems of river classification where individual short stretches of river are classified into differing types, in concurrence with the changing ecological conditions from source to the sea. Should a river such as the Danube be regarded as a single water body or should it be divided into lengths (reaches) corresponding to the underlying geology and physical conditions. If the Danube is subdivided, what about much smaller rivers in other parts of Europe? How long should each reach be? Bearing in mind that each individual water body will require approximately the same amount of effort to investigate and describe it, the number of individual river reaches will influence the amount of overall effort required by individual competent authorities and hence the ability to achieve the overall target of characterisation in the allotted time (by 2004). Because most water administrators in Europe and beyond have collected data on river water quality and flows for many years, as part of their work towards environmental management and pollution control, the availability of additional information, at least for main rivers and larger tributaries is likely to be high. Such information will be required to allow the identification of appropriate ecological conditions, and has indeed already been used for such purposes in many countries.

Lakes

The minimum size of lakes specified by the Directive (0.5 km^2) will lead to the requirement to include many artificial lakes and ponds created for recreational use or as aesthetic features in public parks and private estates. There are many situations where municipal recreational, garden landscape features or even in some cases naturalised effluent treatment lakes have been constructed, either as stand alone bodies of water where none existed before, or by the modification of natural features in the river environment. At present such lakes are often privately owned and are not often sampled as part of the natural environment. There are usually no records of the features of such lakes. The impact of this Directive will be to require a survey of such features in order to include them within the river basin environment and introduce requirements for their management. It is impossible to estimate the likely numbers of such sites. A characterisation exercise to measure the area, depth and identify the underlying geology is the least that is required to produce the list of lakes and other still surface waters. If it is decided that the optional characteristics are required in order to supplement the information under System B (see page 65 for a full explanation), a major survey will be required in most countries over the three-year period 2001–2004. This is unlikely to have been planned or budgeted.

Coastal and transitional waters

Information on transitional and coastal waters is sparse and difficult to obtain. Transitional waters are bodies of surface water in the vicinity of river mouths which are partly saline in character as a result of their proximity to coastal waters but which are substantially influenced by freshwater flows. This may include river mouths and deltas, coastal lagoons, saltwater marshes and wetlands. Member States have a variety of survey methods to identify and classify such waters. In Northern France, for example, for the coastal fringe a system of zoning into homogeneous water bodies is in operation for water management purposes, and to facilitate a global assessment of the water in terms of quality, biology and hydromorphology. The system is based on the identification of three contiguous areas, along a transverse axis from the coast. A *"close field"* where land contributions are very important and there are strong pollution and salinity gradients: an *"intermediate field"* in which such contributions are less obvious but the gradients are less pronounced, and a *"far field"* area in which the land has no effective influence. A study undertaken by the Loire–Bretagne Water Agency indicated that it was possible to break these fields down into discrete areas based on the description of current velocities, the nature of the sea bed and information on biological quality. On the Mediterranean coast morphological characteristics have been used by the organisations concerned with coastal management to determine, within the *"three fields"*, local areas to maintain continuity of management from land to sea (Oudin 2000). The Spanish approach to the identification of transitional and coastal areas is different relying more on local morphology to identify water bodies. Five such bodies are recognised in transitional waters – coastal lagoons, temporal river mouths (in which the mouth is permanently flooded but with high seasonal changes in salinity), coastal marshes, salt pans which are usually of anthropogenic origin, and deep inshore holes heavily influenced by land run-off (Ballesteros 2000). In the UK the coastal zone is defined under a number of Acts of Parliament which do not generally take account of characteristics associated with the water, but are related to navigational and other features of the land mass. The law defines a tidal limit at the downstream boundary of freshwater flow, often associated with a weir or other obstruction and often many kilometres from the coast. The seaward limit of the estuary is defined on a specific map from geographical features. Coastal zones are not formally defined laterally along the coast except in cases where specific legislation applies, such as the Bathing Water Directive, and the outer limit to which water legislation applies is defined as three nautical miles from the low water mark.

The EU Water Framework Directive requires "water bodies" to be identified. Although there is discretion given to Member States on how these can be identified, there is wide variation in the practical means of doing this.

Nevertheless, all countries with coastal environments must arrange to identify individual bodies that can be further characterised in terms of their ecological status using, as the basis, at least the parameters of eco-region, salinity range and mean tidal range.

Groundwater

Precise knowledge of the boundaries of aquifers, and their relevance to river basins, is often unknown. Where countries use groundwater as a source of public water supply, information on their basic characteristics should be available, and in those countries that operate groundwater protection policies, the extent and character of groundwater bodies is usually understood. Maps are available that relate groundwater to geological conditions, but for the purposes of the Directive these must now be related to the overlying river basin topography and activities in river basin catchments. Investigative work is required to ascertain the most appropriate river basin in which to place such aquifers for the purposes of water management.

Characteristics of water bodies

Once identified the natural characteristics of each of the water bodies must be examined in order to identify its "type". Because of their fundamentally differing nature, surface waters and ground waters are dealt with separately here. For surface waters Annex II describes two alternative means of assessing "type" – called System A and System B. Member States may use either system. In the case of heavily modified or artificial surface waters the description must accord with the nearest category – thus, for example, an artificial reservoir might be construed to be a lake, whilst a canal could be considered to be equivalent to a river, or possibly a lake, depending upon its physical nature.

System A for inland surface waters

In order to use this system, Member States must first identify in which "eco-region" each of the water bodies fall. A map has been provided as Annex XI setting out what are regarded by the Commission as distinctive eco-regions. The map is reproduced in Figure 6.1.

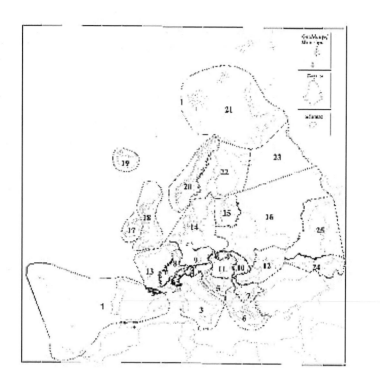

1	Iberic-Macaronesian region	10	The Carpathians	19	Iceland
2	Pyrenees	11	Hungarian lowlands	20	Borealic uplands
3	Italy, Corsica and Malta	12	Pontic province	21	Tundra
4	Alps	13	Western plains	22	Fenno-Scandian Shield
5	Dinaric western Balkan	14	Central plains	23	Taiga
6	Hellenic western Balkan	15	Baltic province	24	The Caucasus
7	Eastern Balkan	16	Eastern plains	25	Caspic depression
8	Western highlands	17	Ireland and Northern Ireland		
9	Central highlands	18	Great Britain		

Figure 6.1 Map showing eco-regions for inland surface water characterisation. (Reproduced, with permission, from the Official Journal of the European Communities (L/327/71; 22/12/2000)).

The eco-regions cross political boundaries and reflect in part the geographical characteristics of the areas and possibly the natural aquatic flora and fauna. The precise relevance to water management of this differentiation is not clear as the variation in water types within each of the eco-regions will usually be larger than differentiation between regions.

Having determined and recorded the basic eco-region within which water body lies, the Member States must differentiate each water body by reference to a number of descriptive parameters. For freshwaters the parameters cover altitude, size of catchment, and a description of the basic geological characteristics in which the water body is located (calcareous, siliceous or organic), and where the water bodies are lakes or reservoirs, depth. Each set of parameters is divided into a set of ranges (for example, the altitude ranges are >800m; 200–800m; and <200m).

System B for inland surface waters

As an alternative to locating the water bodies by reference to the "eco-region" map, Member States may choose to record the precise latitude and longitude, in addition to the altitude, size and geology (and depth, for lakes) of each water body. Paragraph 1.1(iv) of the Annex insists that if System B is used the resulting information must lead to at *least the same degree of differentiation as would be achieved by System A*. The Directive indicates that it may be necessary to use additional descriptors (called optional descriptors in the Annex) to achieve this. This requirement seems to be superfluous as there can surely be no argument that the inclusion of latitude and longitude in place of simply locating the water body somewhere within the area of an eco-region (stretching in several cases over many thousands of square km) achieves and *exceeds* this aim without the need for additional information.

System A for transitional and coastal waters

Transitional and coastal waters have to be placed in the appropriate sea area – the Baltic, Barents, Norwegian, North or Mediterranean Seas or the North Atlantic Ocean, and again a map is provided (in Annex XI) to demarcate the boundaries (particularly between the Barents, Norwegian and North Sea areas) according to the agreed EU decisions. Further typing is achieved by recording the mean annual salinity ranges and the tidal ranges.

System B for transitional and coastal waters

In a similar manner to the procedure for inland waters, the alternative to the use of System A is to record latitude and longitude in addition to the salinity and tidal ranges. The same comments apply to the use of this alternative as for inland surface waters. System B would undoubtedly provide a greater degree of differentiation without recourse to the optional descriptors.

Type specific reference conditions

The purpose of identifying the type of each surface water is so that reference ecological conditions may be identified which would be appropriate to each water body if it were in a pristine state, undisturbed by human intervention. This is so as to provide a baseline of ecological quality against which current conditions and future improvements can be judged.

Simple calculation shows that the possible number of different types of riverine water types in the EU according to the methodology outlined in Annex II is 900 if System A is adopted. System B would supply very many more types. The possible number of System A lake types would amount to some 2700. In the saline environment there could be 90 types of possible ecosystem in estuaries and coastal waters under System A and similarly many more if System B is used. The implications of this system of categorisation is that all waters in the EU will fit into one of the 3780 ecological situations that are identified as baselines for the above categories. Alternatively, if System B is used the degree of categorisation must be much larger with a concurrent extension in the number of possible ecological states.

A number of countries have developed systems to assess and classify their inland and coastal water bodies. In fresh running waters, planktonic algae, periphyton, microphytobenthos, microzoobenthos, aquatic vegetation and fish have all been used as indicators of quality. During the 1970s an intercalibration exercise organised by the Commission demonstrated that the use of benthic macro-invertebrates as water quality indicators appeared to give the most successful results for river assessment. For estuarine and coastal waters the situation is less well developed, the problems compounded by the fact that useful generalisations of such indices will be difficult because of the great natural variability in morphology, climate and tidal regimes and at present there is little evidence that ecological surveys are capable of classifying estuaries in order to assess the ecological status between the effects of nature and anthropogenic influences. This will be a major difficulty if the simplistic approach of the Directive is to be met.

The results from the system adopted could be expressed as an index or score, and could therefore be visualised on a map by means of codes showing levels of quality. There are three main types of such indices; those based on the identification of indicator species that represent a particular level of water quality (sometimes called saprobic indices); those that are based on species diversity as a measure of degradation or deviation from a natural situation (known as diversity indices), and those that combine indicator species with diversity (biotic indices). Such indices may be used comparatively, where reference data is available to compare the actual situation with what should be present. These are called comparative indices. The end result should be an index or score for each water body type in each eco-region, which may then be used as a comparative index to show how much degradation has occurred as the result of human intervention.

Hydromorphological and physico-chemical conditions

In order to assess the baseline ecological characteristics a range of parameters is used to describe the conditions in each individual water body. The characters include an assessment of the basic physical and chemical state of the water and its immediate surroundings (such as the river bed conditions). This is done so as to indicate the conditions under which an ecological regime will have developed. It identifies the basic characteristics that encourage plant and animal life. Whilst the values of the parameters are unique to each site the assumption is made that certain common features will emerge that enable a characteristic type of ecology to flourish if there is no interference by man.

Annex V defines the parameters that should be taken into account in determining ecological characteristics. For each of the water body types that have been identified, type-specific hydromorphological and physico-chemical conditions must be identified that represent the normal undisturbed state of such waters.

The parameters include the flow regime in rivers (quantity and dynamics of flow), river continuity (undisturbed by man and allowing migration of aquatic organisms and sediments), lake levels and residence times (which must represent totally undisturbed conditions), undisturbed morphology, nutrient and other chemical parameters such as dissolved oxygen pH and temperature are within the range normally associated with undisturbed conditions and naturally occurring chemicals are at background levels. Synthetic chemicals must be close to zero or less than the limits presently detectable.

In transitional and coastal waters, the tidal regime must correspond to undisturbed conditions, the depth variations, substrate conditions and structure and condition of inter-tidal zones must be undisturbed, nutrients and other chemical parameters, including transparency, must be within the range normally associated with undisturbed conditions and naturally occurring chemicals must be at background levels. As with freshwater sites, synthetic chemicals must be close to zero or at least less than the limits presently detectable.

Biological reference conditions

The main purpose of the Directive is the safeguarding of ecological status. For all of the waters identified in this survey, type specific biological reference conditions have to be identified as the base-line for action.

Annex V lists the biological organisms that must be taken into account in deriving such conditions. These include for rivers and lakes: phytoplankton, macrophytes and phytobenthos, benthic invertebrate fauna and fish fauna. In the case of transitional waters and coastal waters the biological elements are: phytoplankton, macro-algae, angiosperms, benthic invertebrate fauna and fish; and for coastal waters: phytoplankton, macro-algae, angiosperms and benthic invertebrate fauna. For all of these organisms the definitive level must be that at which the composition and abundance generally represent undisturbed conditions.

The Directive permits the use of spatially based reference conditions or conditions derived from biological modelling work, and where neither of these systems is available then expert judgement may be used to derive values for the biological communities that should exist at each site if these were undisturbed by man. If it is decided to use a spatially based approach to the identification of these conditions, a reference network of undisturbed sites has to be established of sufficient size to establish adequate level of confidence (Annex II (1.3)). This means that the competent bodies must identify some particular sites that conform to the basic physical characteristics of the water bodies in the territory where they exist in undisturbed conditions and identify the organisms growing therein, and the interrelationships between them, to enable a description to be made of an ecological condition that would be considered to be unaffected by man. The assumption is then made that this ecological pattern would be reproducible in all water bodies that conformed to this identified type, if they were also to be in pristine condition.

If modelling is used, instead of observation, to derive the description of pristine ecological status, there must also be a suitable level of confidence applicable to the results to ensure that the values are consistent and valid for each water body type.

Groundwater

The characterisation of groundwaters is carried out in a different way. Groundwaters must be identified by location and boundaries although it is permissible for Member States to group groundwaters together for an initial review. An assessment is needed as to the pressures affecting groundwaters including such issues as pollution sources, abstractions, and artificial recharge regimes, and a view on the general character of overlying strata in the catchment areas. Identification of direct links to surface water ecosystems must also be determined in the initial characterisation exercise.

Where groundwaters are considered to be subject to risk, further work is required to collect more information on the geology, and hydrogeology, and other characteristics of the catchment area.

REVIEW OF THE IMPACT OF HUMAN ACTIVITY

The review of the impact of human activity required by Article 5 is described more fully in Annex II of the Directive. It comprises two activities. The first is an exercise in data collection. The second uses these data to assess impact.

Identification of pressures

Pressures on surface waters

The competent authorities in each river basin district must arrange to collect information on all of those activities that could have an impact on water status. These include the identification of point and diffuse sources of pollution of surface waters, abstraction of water from rivers and lakes which is then used as drinking water, or for industrial or agricultural use. The information must take account of seasonal variations, annual demands and water losses in the distribution network. Details of flow regulations in the river channel and water transfers between rivers or water bodies, significant changes made to the riverine structures, land use patterns and any other human impacts on water status are also required.

Some of this information is already available in the form of returns to the Commission or from the management of implementation programmes for existing EU legislation. The Directive specifically draws attention to the information that is available as a result of the implementation of other environmental instruments such as the Urban Wastewater Treatment Directive, the IPPC Directive, and several the others concerning point sources of pollution;

and the Nitrate Directive, and Plant Pproducts Directive where diffuse sources are concerned.

As far as the practical implementation of this requirement for surface waters is concerned, existing organisations concerned with river management and pollution control will undoubtedly already have significant sources of information readily available. The European Environment Agency in its report on the state of the environment in Europe (EEA 1995) was able to draw on extensive information provided by Member States. Some countries already publish annual reports on the state of the environment in their own countries containing much of the detail needed. The UK government for example (DETR 1998) has published an annual report on environmental statistics for many years, and the Environment Agency now produces its own annual survey of the situation (for example, EA 1998). The information in such reports is generalised, and for the purposes of the Directive, will have to be related to river basins.

Pressures on groundwater

The pressures on ground waters that have been shown to be at risk, and also particularly for bodies of groundwater that cross the borders between two or more Member States, are similarly to be identified. The types of pressures are listed in Annex II, and comprise particularly the location of abstraction points (except those from which less than 10 m^3 per day are withdrawn or those serving less than 50 persons), details of annual average abstraction rates and chemical composition of those waters; location of recharge sites and rates of discharge into them and their chemical composition, and details of land use in recharge areas. The impact of anthropogenic activities on groundwater are largely those related to abstraction of water or connected with the recharge rates, and activities on the surface or discharges into the aquifer that affect the quality of the water in it. In order to assess the impact of human activity of quantity status, information will be required about all abstractions.

Abstraction

Groundwater is commonly considered to be important in arid or semi-arid countries. However, water use data reveal its worldwide importance. The reasons for this include its availability close to where it is required, and its good natural quality, which is usually adequate for potable supplies with little or no treatment, and the low capital costs for the development of boreholes. The ability to phase well development to keep pace with rising demand make it the preferred choice in many situations. In Europe, groundwater plays a major part in the supply of water and many of its major cities of Europe rely on groundwater. Table 6.1 shows the proportion of groundwater in drinking water supplies in some European countries.

Table 6.1 Percentage of drinking water supplied from groundwater in some European countries (EEA, 1999; UN ECE, 1999).

Austria	99	Bulgaria	60
Denmark	98	Finland	57
Hungary	95	France	56
Switzerland	83	Greece	50
Portugal	80	Sweden	49
Slovak Republic	80	Czech Republic	43
Italy	80	United Kingdom	28
Germany	72	Spain	21
Netherlands	68	Norway	13

Groundwater is also used for irrigation, livestock watering and industrial production. It has an important ecological function in those situations where groundwater discharge supports inland and coastal wetlands and the baseflow component of rivers.

The way in which groundwater is abstracted is an important issue to be considered as an impact of human intervention in groundwater status management. In the UK, three large consolidated aquifers provide half or more of the public supplies in the south, centre and east of the country from a few hundred high yielding boreholes. In the UK, groundwater protection policies designate zones from which the recharge is derived around these supplies, in which certain potentially polluting activities are prohibited or controlled. While this is sometimes hydrogeologically problematic, given the complex local groundwater flow systems which are often encountered in these aquifers, it was logistically reasonable to envisage protection zones around a relatively small number of very large abstractions. The situation would be very different in the case where groundwater was drawn from a very large number of much smaller groundwater supplies, such as for example, in Portugal. Because of rapid groundwater flow and response times in karstic aquifers as for example, in parts of Austria, these may be especially vulnerable to pollution. In addition, karst springs may be supported by groundwater recharge from large catchments that are difficult to define. Other examples of groundwater regimes, for example those deriving from the plains of Hungary, which are influenced by the riverine input from the Danube or the Tisza Rivers, require other methods of protection.

Groundwater recharge

Underground fresh water is recharged either by direct infiltration of rainfall, or from rivers and lakes. Following man's interference with the hydrological cycle, recharge can also be derived from canals, reservoirs, irrigated land, water mains and sewerage systems in urban areas, mining waste, sewage lagoons, and other artificial water bodies connected with the subsurface. This means that groundwater recharge is not always of the same good quality as rainfall. Further, information is needed on the relationship between rainfall infiltration and recharge from other aquifers (which may be saline or otherwise contaminated). Part of the rainfall is also lost though evapotranspiration from the soil zone so only a part of the infiltration water contributes to recharging the aquifer. In the subsoil and rock closest to the ground surface, the pore spaces are partly filled with air and partly with water. This is known as the unsaturated or vadose zone and can vary in depth from nothing to tens of metres. The depth of the unsaturated zone is important in the attenuation of pollution.

Land usage and land developments that interfere with groundwater flow, in particular those which affect the ability for water to percolate to the water table, such as the development of hard landscapes as a result of the expansion of built-up areas are important factors in maintaining good water quantity status and gaining information about these and their relationship to the groundwater is an essential part of the river basin survey.

Pollution

In a 1995 study of the potential for groundwater contamination by point sources of pollution in England and Wales (EA 1995) it was found that of the 1205 known point sources of pollution which were examined, 210 directly affected 251 water abstraction points whilst 777 sources of pollution were known to have had some deleterious effect on groundwater quality in aquifers and 425 were suspected to have an adverse effect. Of the 1205 sources it was estimated, using a subjective ranking scheme, that 8% had a highly significant impact, 17% had a medium to high impact, and 28% had a medium impact. This was a limited study carried out to assess possible impact and the study concluded that the true effect of point source pollution was likely to be very high. It is interesting to note that the pollutants included a number of substances that appear on the list of dangerous chemicals in Annex VIII of the Directive and that will therefore by 2013 require action to prevent their discharge to groundwater. These included metals such as arsenic, copper and chromium, solvents, pesticides, various hydrocarbons and cyanide. The problems shown by the study which pertain to the new Directive requirements are that many of the point sources are not amenable to regulatory control in the conventional sense as the sources included landfill sites, with contamination from leachate, industrial sites with no known direct discharges, petrol stations at which contamination was caused by spillage or leaking tanks, dry cleaning plants where solvent contamination of the ground is widespread.

Assessment of impact

Having gathered these data, the competent authorities have to assess the impact that the pressures may have on the status of the waters. In particular the Directive asks the question whether the water bodies will fail to meet their objectives as a result of the pressures. In order to carry out such a survey on a river basin basis for the new Directive a number of questions need to be answered such as:

- What is the geographical extent of pollution?
- What river basins are affected?
- What groundwater remediation has been undertaken or is needed?
- What are the main contaminants?
- What is the severity of pollution?
- What are the main industries and activities that cause pollution?
- What can be done in the way of regulatory or practical control?

ECONOMIC ANALYSIS

Article 5 introduces the concept of economics into the water management regime. An *economic analysis of water use* must be undertaken. Annex III contains the guidance relating to this article. No details are provided as to the means of carrying out such an analysis, but sufficient information must be gathered to take into account the principle of cost recovery for water services and to estimate the volume, prices and costs associated with water service, and relevant investment and to make judgements about the cost effectiveness of possible water use related measures that may be included in the "programme of measures" of Article 11.

The costs of water services are paid for in a variety of ways by different countries. In some cases the services are fully privatised, and the users pay for the service on the basis of costs plus profits to the companies concerned. In other extremes the services are seen as a totally public service and provided by the local or national public administrations, funded by central or local government – the users pay on a variety of bases ranging from cost recovery to a subsidised service. In no cases are the true overall costs of environmental damage caused by the loss of water from the environment, and the damage from returned wastewater, taken into account. One of the purposes of the economic assessment is to permit judgements to be made about the most cost effective combination of water uses to be made in the programme of measures of Article 11. But the relevant paragraph of Article 11 (Paragraph 3 (b)) refers to a basic requirement to observe the conditions of Article 9 concerning the recovery of costs. This article specifies that environmental and resource costs should be taken into account (Article 9(1) – "*have regard to*") and the pricing of water services should *contribute to the environmental objectives* of the Directive.

Whilst countries in the Community will have information on the costs of providing water and sewage infrastructure, the assessment of environmental costs is only now being developed as a practical issue. There is often a number of competing uses for water in a natural water body. Water supply abstraction is one use, but other non-abstractive uses are possibly more important. These may include navigation, aquaculture, power generation, recreation and the non-use category of ecological protection. Full cost pricing will deal with the issue of the funding of water and sewage provision, and will effectively stimulate more efficient resource allocation between competing uses. However, many of the costs related to water abstraction are not captured by such market pricing (whether in a privatised or public costing system). They are external to the market system. Examples might be such issues as over-abstraction having a deleterious effect on wetland habitats and their bird populations for example, or effects of an inter-river transfer changing the local ecology. The valuation of the costs of such externalities as part of the overall cost recovery requirements of the Directive lead to the necessity to use economical valuation techniques that are not normally part of the routine water services cost equations, and this will inevitably be an additional burden on the early implementation activities (McMahon and Moran 2000). Further discussion on economics is given in Chapter 14.

REFERENCES

Ballesteros E., (2000) Defining types in the Mediterranean, *Ecological Status of Transitional and Coastal Waters*, Edinburgh.
DETR (1998) *Digest of Environmental Statistics No 20,* Department of the Environment, The Stationery Office, London.
EA (1995) *Groundwater pollution – evaluation of extent and character from point sources in England and Wales*, Environment Agency, Solihull.
EA (1998) *State of the Environment in England and Wales -Fresh Waters*, The Stationery Office, London.
EEA (1995) *Europe's Environment – The Dobris Assessment*, European Environment Agency, Copenhagen.
McMahon, P., Moran, D., (2000) *Economic Valuation of Water Resources*, CIWEM, London.
Oudin, C.L., (2000) Progress towards a classification system for France, *Ecological Status of Transitional and Coastal Waters*, Edinburgh.

7

Environmental objectives

ASSESSMENT OF WATER STATUS

Article 4 of the Directive asks Member States to use their programmes of measures to prevent deterioration of the ecological status of surface water and aims to achieve good surface water status or, in the case of heavily modified or artificial water, good ecological potential and good chemical status within 15 years (or within an agreed extended timescale). It also aims to achieve good groundwater status over the same timescale. In the case of protected areas, specific standards and objectives may apply, and the Directive requires these to be met also within 15 years. In each case where more than one set of objectives apply, the most stringent one must be used.

The use of common definitions of the status of surface and groundwaters in terms of quality and quantity is an essential part of the process, and environmental objectives are to be set for each water body that define good status and that direct activities towards meeting this. The Commission recognises that there are diverse conditions throughout the Community's area and these may require specific local solutions. A principle of the Directive is thus to take decisions as close to the location of the appropriate water body as is possible. As a result of this view, although common definitions are to be used, and common principles for improvement activities, these may vary in precise nature throughout the Community. This may be seen as a facet of the subsidiarity principle, but it is also a recognition of the wide variation in conditions between the extremities of the European area.

There has been a great deal of discussion on whether it is wise to attempt to devise common European environmental standards for water. The first attempt to do so could be considered to be the standards adopted in the 1978 Freshwater Fish Directive (78/659/EEC). Whilst the preamble to this Directive justified its introduction, at least in part, on the grounds of harmonising existing laws on the grounds of eliminating unequal competition under Article 100 of the Treaty of Rome, the Commission took the view that it was necessary to safeguard freshwater fish from the consequences of pollution in order to achieve the Community's objectives in environmental protection. In the case of this Directive the environmental protection methods involved the adoption of common water quality standards for waters which supported, or were thought to capable of supporting, populations of salmonid or cyprinid fish. Even at this stage the Commission acknowledged that there were differences in the status of waters – they were effectively divided according largely to natural influences into high quality, unpolluted waters where the dominant species were salmon and trout, and those other waters, generally lowland, where the conditions of flow, topography and quality only permitted course fish – pike, perch and so forth – to thrive. Furthermore, the parametric values of zinc and copper were variable depending upon the nature of the water hardness, and Member States were permitted to fix less stringent values for some other parameters such as ammonia and dissolved oxygen, where natural or geographical conditions affected the values.

Alongside such chemical based systems work had been progressing on more general ecological quality criteria in many countries. In 1991 a conference was held in Brussels (Newman *et al.* 1992) to examine the possibility of developing an ecological assessment of river water quality for the management of rivers in order to ensure that a proposed directive on the ecological quality of river waters would have a sound scientific basis.

The use of biological assessment techniques is widespread throughout Europe and beyond, and whilst there are differences in approach, a conclusion of the conference was that there was a surprising similarity between many of the schemes. Most assessment procedures utilise biological indices that are based either on indicator species which are known to exist in and therefore reflect different categories of water quality (known as saprobic indices); or they use systems that have been developed from the principles of the saprobic index which use species diversity as a measure of the degradation of water quality from what might be considered the natural quality (diversity indices); in some cases indices combine the two (known as biotic indices). The majority (approximately 60%) of indices used in European countries fall within the category of Biotic Index whilst the formal Saprobic Index is much less used in Western Europe than in the countries of Central and Eastern Europe. Sampling and identifying indicator organisms which belong to different biotic groups such as algae, macrophytes, invertebrates and fish, provides a means of assessing the changes in water quality and its environment over the long term. In practice the following biological communities have been used as the basis for river water quality assessment:

- Plankton;
- Periphyton;
- Microphytobenthos;
- Macrozoobenthos;
- aquatic macrophytes;
- fish

The saprobity index system is difficult to use, dependant as it is upon expert identification of organisms at species level, and it has largely been overtaken by systems that require less information such as the biotic index. In the UK for example biologically based scoring systems such as the Biological Monitoring Working Party (BMWP) score and the Average Score Per Taxa (ASPT) have been used to assess water quality. According to Pauw (Pauw *et al.* 1992) in seven of the EU countries (Belgium, Denmark, France, Germany, Ireland, Luxembourg and the UK) biotic index methods have been accepted as a national standards, whilst in other States such methods are used at least as regional standards.

A drawback of some of these systems is that they do not always take account of the natural environmental conditions that may influence biological growth populations, such as water flow, temperature, altitude, and the natural chemistry of the area. Computerised models have been developed that take such issues into account and predict the expected type of flora and fauna, and such systems may offer potential to develop criteria reflecting the natural state of waters. This could be particularly valuable for transboundary waters because they take into account the properties of river catchments that influence their ecology. In the UK a model known as RIVPACs has allowed the prediction to be made of the invertebrate communities that would be expected to be found in unpolluted rivers in certain geographic areas, by taking the physical and chemical features of the area into the equation (Chave 1993). By examining the theoretical picture with the results of surveys, it is possible to prepare an index (known as the ecological quality index) which represents the ability of the river to sustain an appropriate invertebrate community and by implication provides a method of assessing the ecological status of a particular river reach. Such a system could be used to develop standards that could apply across the country and beyond.

In Belgium a biotic index using macro-invertebrates has been developed using as its basis indices which originated in the UK and France. Standardised techniques of sampling and the identification of organisms is carried out and the resultant index values derived from these data are divided into five water quality classes ranging from unpolluted waters to very heavily polluted. This system has been used since 1979 to prepare coloured maps indicating water quality. The system has been adopted in the EU by Spain and Portugal and is also used outside the EU. The index provides information on the effects of organic and inorganic pollution, physical changes in the riverine environment, and the effectiveness of the self-purification of watercourses. It has been used for water quality planning purposes as it is capable of showing the effectiveness of anti-pollution measures, and can be used to identify suitable sites for wastewater treatment plants and similar structures.

DIRECTIVE SPECIFICATIONS FOR WATER STATUS

Protected areas

Although the Directive aims to achieve a condition known as good ecological status in its surface waters (and good water status in groundwaters), numerical standards are only set at Community level for waters that are designated for protection under existing EU legislation – the *protected areas* of Article 6. Such standards are generally of a chemical nature, and are contained, generally, in the annexes of existing directives, many of which will be absorbed within the Framework Directive. Annex IV specifies the relevant directives. For natural surface freshwaters rivers and lakes these are:

• *Directive 75/440/EEC concerning the quality required of surface water intended for the abstraction of drinking water*, which specifies limits for abstracted water in three categories depending upon the degree of treatment plant installed;

• *Directive 78/659/EEC on the quality of fresh waters needing protection or improvement in order to support fish life*;

• *Directive 91/676 on the protection of waters against agricultural pollution from agricultural sources*, which sets out a nitrate limit of 50 mg/l in surface waters;

• *Directive 76/464/EEC on pollution caused by certain dangerous substances discharges into the aquatic environment of the Community*, and its five "daughter" directives, each of which sets environmental quality standards and emission limits for the substances concerned;

• *Directive 91/271/EEC concerning urban wastewater treatment*, which specifies treatment according to the degree of eutrophication of the water receiving the effluent; and

• *Directive 76/160/EEC concerning the quality required of bathing water*, which applies to chemical and bacteriological standards to inland and coastal recreational waters.

Table 7.1 The key parameters in directives as specified in Annex IV for natural surface freshwaters rivers and lakes.

Parameter	Abstraction			Freshwater Fish		Nitrate	Dangerous Substances	Urban Wastewater	
	A1	A2	A3	Salm-onid	Cypr-inid			Sensitive waters	Works
PH	6.5-8.5	5.5-9	5.5-9	6-9	6-9				
O_2	>70%	>50%	>30%	50%>9mg/l	50%>7mg/l				
As µg/l	50	50	100						
Cd µg/l	5	5	5				5		
Cr µg/l	50	50	50						
Cu µg/l	50	100							
Hg µg/l	1	1	1				1		
Ni µg/l									
Pb µg/l	50	50	50						
Zn µg/l	3000	5000	5000	<300	<1000				
NH_4 mg/l		1.5	4	<1	<1				
NO_3 mg/l	50	50	50			50		50	15
NO_2 mg/l									
PO_4 mg/l									2
Pet Hydrocarb mg/l	0.05	0.5	1.0	3	3				
PAH µg/l	0.2	0.2	1.0						
CCl4 µg/l							12		
pp-DDT µg/l							10		
Total DDT µg/l							25		
Pentachloro-Phenol µg/l							2		
Aldrin ng/l							10		
Dieldrin ng/l							10		
Endrin ng/l							5		
Isodrin ng/l							5		
Hexachloro-Benzene µg/l							0.03		
Hexachloro-Butadiene µg/l							0.1		
1,2-dichloro-ethane µg/l							10		
Trichloro-ethylene µg/l							10		
Perchloro-ethylene µg/l							10		
Trichloro-Benzene µg/l							0.4		
Hexachloro-Cyclobenzene µg/l							0.1		

The Bathing Water Directive and the Urban Wastewater Treatment Directive apply to transitional and coastal waters, and a further Directive *79/923/EEC on the quality required of shellfish waters* is also relevant.

Table 7.2 Key parameters in the Bathing Water and Shellfish Directives.

Parameters	Bathing waters	Shellfish
PH	6-9	7-9
Total coliforms	10,000	
Faecal coliforms	2,000	
Salmonella	0	
Enteroviruses	0	
Colour	No abnormal change	No more than 10 mg Pt/l from background
Mineral oils	No visible film	No visible film
Detergents	No lasting foam	
Phenols	No specific odour	
Transparency	1m	Not more than 30% more than normal
Salinity		Not more than 40ppt
Dissolved oxygen		Not less than 70%
Organohalogens		No harmful effect on shellfish
Metals		No harmful effects on shellfish
Taste		No impairment of taste

The standards in these directives are generally based on chemical criteria, with some bacteriological components, although the EU Water Framework Directive itself refers to "good water quality" which includes an assessment of the state of the watercourse in terms of the aquatic biota and other biological criteria. A number of countries have produced criteria for assessing the current state of a river reach when compared with the expected pristine state at that point given the rate of flow and other physical characteristics of the watercourse. The saprobic index and biodiversity index are examples of these.

Good water status for surface waters

If we now turn to the Directive definition of good water status, this is set out in Annex V. The Directive requires Member States to assess the status of four categories of natural waters through the observation of a number of parameters which include biological elements, the physical and quantity characteristics of the water and its surroundings, and the physico-chemical condition of the water.

Table 7.3 Parameters to be addressed in determining surface water status.

	Rivers	Lakes	Transitional waters	Coastal waters
Biology	Aquatic flora Benthic invertebrates Fish	Phytoplankton Other aquatic flora Benthic invertebrates Fish	Phytoplankton Other aquatic flora Benthic invertebrates Fish	Phytoplankton Other aquatic flora Benthic invertebrates
Hydro-Morphology	Water flow regime Connection with aquifers River continuity Depth, Width River bed conditions Riparian zone	Water flow regime Connection with aquifers Residence time Depth Lake bed conditions Lake shore	Depth variation Bed conditions Inter-tidal zone Freshwater flow Wave exposure	Depth variation Coastal bed Conditions Inter-tidal zone Current direction Wave exposure
Chemistry	Thermal conditions Oxygen levels Salinity Acidification status Nutrients Pollutants in discharges Priority substances	Transparency Thermal conditions Oxygen levels Salinity Acidification status Nutrients Pollutants in discharges Priority substances	Transparency Thermal conditions Oxygen levels Salinity Nutrients Pollutants in discharges Priority substances	Transparency Thermal conditions Oxygen levels Salinity Nutrients Pollutants in discharges Priority substances

In order to assess whether the surface water falls within the appropriate category of "good water status" Member States will have to carry out sufficient monitoring of all of these characteristics as a minimum requirement, and then use the data to place the water body in one of three classes: high status, good status; and moderate status. Annex V sets out general criteria that may be used to make the judgements. There are no numerical values in the classification system, because the classifications are designed to allow for regional variations in natural conditions.

For rivers, lakes, coasts, transitional waters and coastal waters there is a general definition of high, good and moderate status. The definitions relate to the amount of deviation from undisturbed conditions as a result of human intervention. Thus "High status" would be waters in which there are no, or only very minor changes in the natural physico-chemical and hydro-morphological conditions, the biological quality elements represent the substantially undisturbed situation and the waters correspond to the type specific conditions described in Chapter 6.

"Good status", the status which is the principal target situation as a result of implementing the Directive, is described generally as a situation where the values of the biological elements for that type of water show evidence of the impact of human activities, but *deviate only slightly from those normally associated with the surface water body type under undisturbed conditions.*

Waters of "moderate status" deviate moderately in biological quality from the conditions which should occur under an undisturbed state. There is no formal definition of the word moderate.

There are two further status definitions for surface waters – poor and bad. Poor waters are those in which major alterations in biological quality and biological communities from an undisturbed state are evident; and bad waters are those in which large deviations are observed, and large portions of the normal biota are missing.

Table 1.2.1 of Annex V describes in more detail the values which must be adopted under each of the quality status definitions for rivers.

Ecological quality ratios

One of the problems inherent in the application of a variety of classification schemes by individual Member States is the likelihood that differing interpretations will be laced on the results of assessments, leading ultimately to different levels of ecological quality being classified as the same. Thus in one region the particular ecological status may be considered "good" on a set of data whereas a similar but not necessarily identical set of data might lead to the classification as "moderate". The inexactitude of biological responses to outside influences (both anthropogenic and natural) on a biotic system will exacerbate this likelihood. In order to attempt to overcome the problem the Commission is intending to use standard methods of analysis, so that errors in the data itself are minimised, and the acceptable methods are listed in Annex V (1.3.6). Second, the results of the systems operated by each Member State must be expressed as "ecological quality ratios". The ratios will be the relationship between the observed biological parameters in a water body, (for example a numerical value of the number of taxa or similar measurement) compared with the expected numerical value for the pristine reference condition for the type of water body. The ratio is expressed as a numerical value between one and zero. High ecological quality would thus be a value of one, and poor ecological quality would equate to a zero value. The classification of waters is achieved by dividing the ecological scale of 0–1 into five classes ranging from high to bad ecological status. Figure 7.1 shows the resultant scheme.

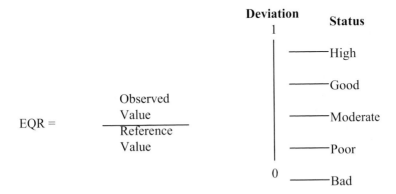

$$EQR = \frac{\text{Observed Value}}{\text{Reference Value}}$$

Figure 7.1 Ecological Quality Status outline

In order to bring about consistency in the ranking the Commission will set up an intercalibration exercise in which Member States will identify a range of sites to covering the each eco-region. Specific characteristics will be needed at each site to ensure that at least two of the sites in each eco-region are close to the upper and lower boundaries of what will be the definitive "high", "good" and "moderate" ecological status. Once the sites are identified each Member State will apply its own classification assessment to the identified sites and to a surface water body of same kind in its area, and the resulting numerical values will be used to set class boundaries for general use. The timescale for this exercise is extremely short, and Member States and the Commission have set a task which for certain types of water body is almost impossible to achieve. If it is carried out badly, the resulting classifications may have major implications for investment decisions, and could prove to be a costly mistake for the future, bearing in mind the lack of experience in using many of the parameters that are specified in Annex V. The Member States have until December 2003 to identify bodies of water with the appropriate characteristics, and the Commission has to publish a consolidated list by 2004. The intercalibration exercise must be completed by June 2005 and published by December 2005.

EXISTING SURFACE WATER CLASSIFICATION SYSTEMS

River classification

How does the classification system for rivers compare with what is in place in Member States and other accessing countries at the present time, and will this have an impact upon monitoring and assessment programmes already in place? Most of the countries which utilise a classification system for their surface waters divide their ranges of quality into three, four or five classes. There is a predominance of five-class systems. Examples of these are found in the UK, Belgium and many of the Eastern European nations. In the UK, a "River Ecosystem" (RE) classification has been developed in response to a legal requirement under its Water Resources Act. This is based purely on river chemistry, however, and does not take into account many of the parameters that are identified in the Directive. This is despite a great deal of development work on the RIVPAKs system. The basic standards are set out in a statutory instrument and comprise five classes ranging from high quality water (category RE1) to poor quality (category RE5). The key parameters (Table 7.4) are the dissolved oxygen concentration, the five-day biochemical oxygen demand (BOD$_5$) (using the method of allylthiourea suppression) and the value of total ammonia expressed as nitrogen. Other parameters included in the classification, which are not shown in the table below, are un-ionised ammonia, pH values and levels of copper and zinc which are related to the hardness of the water.

Table 7.4 Values of dissolved oxygen, BOD5 and ammonia in the UK surface water classification scheme

Water Category	Dissolved oxygen 10 percentile values of % saturation	BOD5(ATU) Mg/l, 90 percentile value	Total ammonia Mg/l as N, 90 percentile value
RE1	80	2.5	0.25
RE2	70	4.0	0.6
RE3	60	6.0	1.3
RE4	50	8.0	2.5
RE5	20	15.0	9.0

Biological assessment is also used widely. Macro-invertebrates are the preferred species for assessment because amongst other characteristics they are present in all UK rivers, they are easy to sample, and they represent several different trophic levels so that suitable differentiation of water quality can be made. The use of the BMWP scoring system is common as well as the ASPT system. Water may be placed in five bands of biological quality using these systems. However, they have no legal basis at the present time.

In Belgium on the other hand, a biotic index and a chemical classification is used. The biotic index is derived from sampling of micro-invertebrates and assessing quality on the basis of the observed fauna groups. The index leads to a five level classification system ranging from lightly or unpolluted waters to very heavily polluted. The classification system is shown in Table 7.5 and is used as the formal assessment method for rivers.

Table 7.5 Belgian biotic index values and river water classification

Class	Biotic Index	Significance
I	10–9	Lightly or unpolluted
II	8–7	Slightly polluted
III	6–5	Moderately polluted, critical situation
IV	4–3	Heavily polluted
V	2–0	Very heavily polluted

In addition to the biological classification some investigators use a "Chemical Index" based upon the same key factors as the UK system, namely dissolved oxygen, BOD5 and ammonia, although individual values given to the parameters differ in detail (De Brabander 1992).

In Ireland a quality rating system associates the relative abundance of five key groups of macro-invertebrates to water quality. The system uses five classes ranging from bad quality to good quality (known as Q-values) and several intermediate classes. In order to arrive at the appropriate classification, macro-invertebrate samples are taken by experienced biologists, using approved techniques, and the organisms present are identified and placed into indicator groups representing sensitive organisms, less sensitive forms, relatively tolerant forms, tolerant and most tolerant forms. Further input of other factors such as the presence and abundance of macrophytes and algae, the occurrence of slime growths, and an overall assessment of the nature of the site is made in order to come to a decision on its class. For national reports the resultant Q-values are grouped into four classes: A1– unpolluted, B1– slightly polluted, B2– moderately polluted and C– seriously polluted.

The above represent just a few of the many systems in use throughout the Community at the present time.

Central and Eastern European classification systems

In the accessing nations of Eastern Europe, classification systems are also in use. Slovakia, the Czech Republic and Hungary use a numerical classification system for the biological assessment of water quality using macrobenthos and based on a so-called saprobic index. This reflects the numbers of individual species present in a sample together with a saprobic index for each species, which is related to the tolerance to organic load. The classes, five in number, represent ranges of quality from good quality, non-polluted water to poor quality, very strongly polluted water. The class limits are basically similar in all three countries. Differences may relate to the values placed on certain organisms, which reflect the local physical conditions in the watercourses. A chemical system is also in use.

Table 7.6 Biological classification in the Slovak, Czech and Hungarian Republics

Class	Slovakia Upper Limits	Czech Republic Upper Limits	Hungary Upper Limits
I	1.2	1.5	1.8
II	2.2	2.2	2.3
III	3.2	3	2.8
IV	3.7	3.5	3.3
V	>3.7	>3.5	>3.3

Table 7.7 Chemical classification used by the Slovak Republic

Category: Parameter	I	II	III	IV	V
PH	6.0–8.5	6.0–8.5	6.0–8.5	5.5–9.0	<5.5–>9.0
DO mg/l	>7	>6	>5	>3	<3
COD mg/l	15	25	35	55	>55
TOC mg/l	5	8	11	17	>17
Amm-N mg/l	0.3	0.5	1.5	5.0	>5.0
Nitrite-N mg/l	0.002	0.005	0.02	0.05	>0.05
Nitrate-N mg/l	1.0	3.4	7.0	11.0	>11.0
Tot-P mg/l	0.03	0.15	0.4	1.0	>1.0
Hg µg/l	0.1	0.2	0.5	1.0	>1.0
Cd µg/l	3	5	10	20	>20
Pb µg/l	10	20	50	100	>100
As µg/l	10	20	50	100	>100
Cu µg/l	20	50	100	200	>200
Cr µg/l	5	10	20	50	>50
Ni µg/l	20	50	100	200	>200
Zn µg/l	20	50	100	500	>500

The Hungarian chemical water quality classification is based on five classes in a similar manner to that of Slovakia. In addition to surface water quality parameters and limit values, the Hungarian quality assessment system refers to a number of quality objectives for sediments for toxic metals and trace organic contaminants. Key determinands include the following:

Table 7.8 Hungarian quality standards for surface waters

Class:		I	II	III	IV	V
Parameters	**units**					
PH		6.5–8.0	8.0–8.5	8.5–9.0	9.0–9.5	>9.5
				6.0–6.5	5.5–6.0	
DO	Mg/l	7	6	4	3	<3
COD	Mg/l	5	8	15	20	>20
TOC	Mg/l	3	5	10	20	>20
Amm-N	Mg/l	0.2	0.5	1.0	2.0	>2.0
Nitrite-N	Mg/l	0.01	0.03	0.1	0,3	>0.3
Nitrate-N	Mg/l	1	5	10	25	>25
Tot-P	µg/l	100	200	400	1000	>1000
PO4-P	µg/l	50	100	200	500	>500
Al	µg/l	20	50	200	500	>500
As	µg/l	10	20	50	100	>100
Zn	µg/l	50	75	100	300	>300
Hg	µg/l	0.1	0.2	0.5	1	>1
Cd	µg/l	0.5	1	2	5	>5
Cr	µg/l	10	20	50	100	>100
Ni	µg/l	15	30	50	200	>200
Pb	µg/l	5	20	50	100	>100
Cu	µg/l	5	10	50	100	>100
Hydrocarb	µg/l	20	50	100	250	>250

In Hungary sediment standards are also set out as target values and these are used in conjunction with water quality data and biological assessment as a means of assessing the overall quality of the environment at particular locations. The guidelines make reference to Hungarian soil quality criteria, and use as reference value concentration limits common to several other countries including The Netherlands, Canada and internationally agreed criteria for the River Rhine. The Hungarian soil guidelines are as follows:

Table 7.9 Hungarian soil guidelines

Contaminants	µg/g
Arsenic	15
Cadmium	5
Chromium	100
Copper	100
Lead	100
Mercury	2
Nickel	100
Zinc	300

Moving even further east, the Romanian system is based on three classes. The major chemical determinands are shown in Table 7.10.

Table 7.10 Romanian chemical classification system for surface waters

Determinand	ClassI	Class II	Class III
Ammonium mg/l	1	3	10
Nitrate	10	30	-
Nitrite	1	3	-
Phosphorus	0.1	0.1	0.1
DO	6	5	4
COD	10	20	30
Pet Hydrocarbons		0.1	
As		0.01	
Cd		0.001	
Cr (III)		0.5	
Cr (VI)		0.05	
Cu		0.05	
Hg		0.001	
Ni		0.1	
Pb		0.05	
Zn		0.01	
Triazine and other individual herbicides		0.001	
Organochlorine insecticides		0.0001	
PAH		0.0002	

In addition to these general classifications, as much of the water quality is related to the River Danube, the following targets for the surface waters in the Danube river basin have been suggested to prevent eutrophication. Although eutrophication is said to be a problem in the Danube river basin, levels of nitrate and phosphate would not support this view for this particular river if the EU standards are adopted.

Table 7.11 Targets for the Danube.

Total N mg/l	<2.2 as an annual average
Total P mg/l	<0.2 as an annual average
N:P ratio	10:1

For protection of the Black sea long-term and medium-term targets have also been proposed.

Table 7.12 Target water quality to protect the Black Sea.

Long term	Total N mg/l	0.5
	Total P mg/l	0.1
	N:P ratio	5:1
Medium term	Total N mg/l	1.3
	Total P mg/l	0.13
	N:P ratio	10:1

Examination of the N and P results for some of the tributary rivers of the Danube, for example, the River Morava catchment, indicate that using these criteria the surface waters would be classified under this system as eutrophic, and would contribute a significant input of nitrogen and phosphorus towards exceeding the long- and medium-term values for protection of the Black Sea.

UNECE standard statistical classification of surface freshwater quality

The United Nations Economic Commission for Europe (UNECE) has recently adopted a standard water quality classification based on the requirements of protecting aquatic life (WHO 1997). The limits for parameters in this system tend to be more stringent than for other possible uses of the water. The system has been applied to a number of transboundary waters and may be used under the terms of the Convention on the Protection and Use of Transboundary Watercourses and International Lakes (UNECE 1992). Table 7.13 shows the classification system.

Table 7.13 ECE standard statistical classification of surface freshwaters

Parameter	Class I	Class II	Class III	Class IV	Class V
DO% epilimnion	90–110	70–90 or 110–120	50–70 or 120–130	30–50 or 130–150	<30 or >150
DO% hypolimnion	90–70	70–50	50–30	30–10	<10
DO% unstratified water	90–70	70–50	50–30 or 120–130	30–10 or 130–150	<10 or >150
DO mg/l	>7	7–6	6–4	4–3	<3
COD–Mn mgO$_2$/l	<3	3–10	10–20	20–30	>30
Total P μ/l	<10 (<15)	10–25 (15–40)	25–50 (40–75)	50–125 (75–190)	>125 (>190)
Total N μ/l	<300	300–750	750–1,500	1,500–2,500	>2,500
Chlorophyll–a μ/l	<2.5(<4)	2.5–10 (4–15)	10–30 (15–45)	30–110 (45–165)	>110 (>165)
pH	9.0–6.5	6.5–6.3	6.3–6.0	6.0–5.3	<5.3
Alkalinity mg CaCO$_3$/l	>200	200–100	100–20	20–10	<10
Aluminium μ/l	<1.6	1.6–3.2	3.2–5	5–75	>75
Arsenic μ/l	<10	10–100	100–190	190–360	>360
Cadmium μ/l	<0.07	0.07–0.53	0.53–1.1	1.1–3.9	>3.9
Chromium μ/l	<1	1–6	6–11	11–16	>16
Copper μ/l	<2	2–7	7–12	12–18	>18
Lead μ/l	<0.1	0.1–1.6	1.6–3.2	3.2–82	>82
Mercury μ/l	<0.003	0.003–0.007	0.007–0.012	0.012–1,400	>1,400
Zinc μ/l	<45	45–77	77–110	110–120	>120
Dieldrin μ/l	–	–	<0.0019	0.0019–2.5	>2.5
DDT μ/l	–	–	<0.001	0.001–1.1	>1.1
Endrin μ/l	–	–	<0.0023	0.0023–0.18	>0.18
Heptachlor μ/l	–	–	<0.0038	0.0038–0.52	>0.52
Lindane μ/l	–	–	<0.08	0.08–2.0	>2.0
Pentachlorophenol μ/l	–	–	<13	13–20	>20
PCBs μ/l	–	–	<0.14	0.14–2.0	>2.0
Gross alpha activity mBq/l	<50	50–100	100–500	500–2,500	>2,500
Gross beta activity mBq/l	<200	200–500	500–1,000	1,000–2,500	>2,500

Data in brackets refer to flowing waters; Cd, Cu, Pb, Hg, Ni and Zn levels should apply to waters of hardness from 0.5 to 8 meq/l

Lake water classification systems

The Directive requires objectives to be set for lakes, both natural and artificial. For the purpose of the Directive lake means *a body of inland standing water*. The size definition in Annex II ranges from 0.5 km^2 to lakes of greater than 100 km^2 in area. It is thus likely that all but the smallest ponds will be included as water bodies under the protection of the Directive. Lakes are sometimes natural formations, for example the lakes of the Lake District in England, or the many thousands of small lakes in Finland, ranging in size from small water bodies up to the largest natural lake in Europe, Lake Balaton in Hungary. Beyond natural lakes, water supply has traditionally been provided throughout the world by the construction of large reservoirs. The reservoir sizes range from tiny concrete structures to the huge reservoir at Keilder, Northumberland (UK). A third group of lakes is that constructed for power generation. Again, some of these lakes are immense structures, but there are also dammed rivers and canals within the river networks that contain standing water rather than running water and that therefore may fall within the "lakes" category. Besides the natural water supply or power generation lakes, there are innumerable privately owned natural and artificial lakes and standing water bodies constructed for a variety of purposes including fish rearing and sport, and for architectural and recreational purposes. These are largely unknown at the present time. The number of lakes which will have to be included within the survey of river basins under Article 5, and to require the identification and setting of quality objectives, runs into many thousands. To simply identify and classify by size alone will be a monumental task.

For each lake so identified, quality objectives must be set. The terminology within Annex V is similar to that used for river waters. The set of parameters given in Table 7.3 must be assessed against the general criteria of Annex V and the specific quality classification of Table 1.2.2 in Annex V. Thus each of the biological, hydromorphological and chemical criteria must be judged against the theoretically undisturbed state of the lake and the lake objective set at the "*good*" status level. A complication may become evident in decisions on the status of lakes. Although there are, as indicated above, many natural lakes, and the classification of good quality is a matter of assessing the condition when compared with the undisturbed natural ecology quality, a vast number of lakes are man-made. The Directive requires good ecological potential to be the appropriate quality objective. Many artificial lakes are subject to physical management regimes, such as drawdown for water supply purposes or hydropower generation, or for flood protection. Such management activities are sometimes regular occurrences, but in many cases they are sporadic depending upon outside influences. At times when activities are minimal, the ecology may revert to a near natural state, indeed on some very old reservoirs, it is difficult to appreciate that they are man-made structures, yet the original purpose must not be forgotten, and it is always possible that the operational needs of the construction may have to be put before the achievement of good status. The question of when, or whether, such structures should be regarded as "near natural" may arise.

In practice there are few examples of experience in dealing with the parameters thought appropriate to lake water assessment that are identified in the Directive. This is likely to be an entirely new area of work for the competent bodies set up under the Directive and it seems by no means certain that this aspect will be adequately covered in the time available.

Most existing experience of lake management considers the question of eutrophic state and, in the case of reservoirs, the level of oxygen and other chemical parameters that influence water treatment processes. In some cases, where the lakes are operated for fishery purposes, fish stocks are measured and in most cases positively maintained by introduction of new stock as necessary.

Probably the most widely used classification scheme is that based on an assessment of the eutrophic state using the OECD definitions of eutrophication developed in 1982 (OECD 1982). In this system, lakes are classified, according to their phosphorus and chlorophyll levels, into five categories dependent upon the nutrient status and the amount of algal activity that is taking place within the water body. Table 7.14 sets out the definitions.

Table 7.14 OECD definitions of eutrophic state

Type	Phosphorus mg/m^3	Chlorophyll-a mg/m^3
Ultra-oligotrophic	<4	<1
Oligotrophic	<10	<2.5
Mesotrophic	10 – 35	2.5 – 8
Eutrophic	35 – 100	8 – 25
Hyper-eutrophic	>100	>25

This system has largely come into use because of the perceived and actual problems caused by the move from a low nutrient state to a high nutrient state in the context particularly of the provision of water supplies, but it has also been useful where natural lakes have moved towards a eutrophic state as a result of man's activities such as farming or the discharge of effluent into feeder streams.

The problems of eutrophication from a water supplier's viewpoint are that, as algal activity increases, so the amount and species of algae move towards the growth of toxic blue–green cyanobacterial species rendering the water unfit for consumption, filter blockage occurs, and high levels of iron and manganese may arise. There is a risk of chlorinated by-products being formed and detected in the treated water, and taste and odour molecules may form. As well as the use for water supply, most water supply companies also use their reservoirs as recreational sites, and eutrophication may affect this because fish species may change and spawning grounds be eliminated, the problem of toxic blue–green algae may impact upon recreational use both from the possible toxic effects and there may be a reduction in the attractiveness of the water due to algal scums. Such problems affect the economics of running the reservoirs by reducing income from the activities. There has been much work undertaken in various parts of the world to investigate eutrophication and its problems in lakes, but the Directive requirements appear to change the direction of emphasis in its quality objectives. Routine monitoring of lakes and standing water bodies has been much less extensive than is the case for rivers. Most data have been derived as a result of the occurrence of specific problems, such as the onset of blue–green algal blooms or as a result of specific water treatment problems. There are numerous data-sets available for specific lakes as a result of long term studies carried out by research organisations to assess the changing eutrophic state and to examine algal growth regimes – for example, in the UK, studies have been carried out in the Lake District and the Norfolk Broads over many years. The Lakes, which are part of the Environmental Change Network, have also been subjected to long-term monitoring of algae, macro-invertebrates, macrophytes, diatoms and nutrients. However, on the whole, data as required by the Directive is scarce and the application of objectives of the type specified is a largely unknown issue.

Objectives for transitional and coastal waters

The setting of objectives for transitional and coastal waters will be a complex issue in many countries as these waters may cover an extensive range of types. Transitional waters include the estuaries of rivers, coastal wetlands, and salt marshes and coastal lagoons. As with the other classification schemes, the Directive aims to achieve "good" status. This is defined as being the status in which biological parameters differ only slightly from the reference values for the water body type and when the environmental quality standards for chemicals are met. Member States must develop a suitable biological classification system for transitional waters using the parameters in Annex V. It is intended that the biological element of the classification systems will be harmonised in due course by the Commission through the examination of the EQRs for the different schemes adopted by Member States.

There is a wide variety in the amount of monitoring and classification work undertaken throughout the Community at present. Some of this is in response to the existence of international agreements. In France for example a system for the evaluation of quality has been developed for all waters including transitional and coastal waters. This utilises a five-band class system based on the suitability for water uses and the maintenance of aquatic life. It uses the set of indicator parameters shown in Table 7.15. These data are then combined with information on water uses to reach a decision on its ecological quality class.

Table 7.15 Quality indicators in the French system

Indicator	Parameters
Temperature	Temperature
Salinity	Conductivity
Suspended solids	Turbidity, transparency
Oxygen	
Macro-waste	Anthropogenic/natural
pH	Water pH, redox potential – sediments
Nutrients	N,P in water and sediments
Organic matter	C, N in water and sediments, N/P ratios
Microbiology	In water, sediments, organisms
Metallic micropollutants	In water, sediments, organisms
Organic micropollutants	In water, sediments, organisms
Radioelements	In water, sediments, organisms
Non-toxic algae	Chlorophyll-a
Toxic algae	Toxicity tests
Floating hydrocarbons	On surface layer

In Italy, the coastal waters assessment scheme uses data on the chemical and physical status of water and sediments, the impact on biota and the accumulation of micropollutants in order to classify such waters. Some countries have carried out a great deal of work in estuarine and coastal waters and have developed very sophisticated techniques.

In the UK two schemes are currently in use. In England and Wales, a classification system was developed by the former National Water Council (NWC) and used from 1980 onwards. The scheme took account of the biological quality as judged by reference to the passage of migratory fish and fish populations, the diversity and quantity of benthic organisms, and the absence of toxic or tainting substances. It also classified waters into four aesthetic categories and assessed dissolved oxygen levels. Each of the three of factors: biological, aesthetic and chemical quality was given a score out of ten, and the sum of the scores added up to a figure which allowed the waters to be placed into one of four classifications. A significant subjective assessment was allowed because of a lack of data. The classifications are as in Table 7.16.

Table 7.16 Estuary classification system (England and Wales)

Class	Number of points	Description
Class A	30–24	Good quality
Class B	23–16	Fair quality
Class C	15–9	Poor quality
Class D	8–0	Bad quality

In 1990, Scotland developed a coastal waters classification scheme to replace the NWC scheme, and in 1999 a variant of this scheme (known as the Scottish Environmental Protection Agency (SEPA) scheme) was also adopted by the Environment Agency for England and Wales. This scheme comprises two parts, for estuarine waters and coastal waters. Both comprise four classes, A to D ranging from excellent to seriously polluted. Unlike the previous system, instead of using a summation of scores, the new scheme uses a default situation in the same manner as the Directive. For each of the four classes, the identified water body (area of estuary or coastal water) is assessed against a set of criteria, and the classification is confirmed by placing the water in the highest class to which all of its condition criteria conform. For the estuarine classification scheme the criteria consist of an appraisal of aesthetic condition; an assessment of fish migration, such as salmon or eels; an assessment of the condition of the resident fish community; an examination of the resident biota which should be consistent with the physical and hydrographical conditions, and/or a sediment bioassay using the amphipod *Corophium sp*; and an assessment of the presence or absence of persistent toxic substances judged against a national background level. Water chemistry is taken into account with specific reference to dissolved oxygen levels and the presence of internationally agreed dangerous substances of the UK Red List and the Dangerous Substances Directive. The coastal waters scheme uses a similar set of appraisal conditions but with the addition of an assessment of bacteriological quality where coastal bathing takes place, in which case the standards of the Bathing Water Directive are applied. There is no reference to fish in the coastal scheme, the biological assessment relies upon an assessment of whether or not the flora and fauna are consistent with the local physical and hydrographical conditions.

Some countries have been very active in developing coastal classification schemes. Norway, for example has investigated conditions in its fjords for many years and introduced in 1993–94 a system for classification of environmental quality, degree of pollution and suitability for various uses based upon measurements of water quality, metals and organic micropollutants in sediments and biota and soft bottom fauna. The scheme uses five classes ranging from very good to very bad quality, involving the assessment of nutrients (nitrogen, phosphorus, chlorophyll-a and Secchi depth, with some supporting parameters such as oxygen, salinity, temperature and phytoplankton species identification); metals in water; sediments and organic contaminants in sediments; and an analysis of organic toxicants in a range of organisms including mussels, fish and crab. The organic content of bottom sediments is also measured in this scheme. Finally an estimation of the suitability for use based primarily on bacterial standards required for bathing is undertaken. Within the Norwegian sphere several monitoring programmes have been undertaking phytoplankton and macrophyte sampling, and the amount of work involved in obtaining reliable data is considerable. Taxonomic work is labour intensive and time consuming. In order to characterise the Skaggerak region, samples are taken from 0 to 30m depth on a fortnightly basis, and the results for ten years work are thought to be necessary to provide a proper basis for describing the natural variability of algae with acceptable confidence and precision. The short time available for the exercise required by the Directive to characterise pristine reference areas of transitional and coastal waters is likely to prove problematic.

The Norwegian work has pinpointed one worrying further aspect of the classification concept that may not be capable of resolution. The Directive principle is to identify areas of high quality, which are well away from human intervention as a means of classifying "high" and "good" status. In one such area, sea urchins have destroyed the normal kelp forests, and this has persisted for 10 years. The biological status is thus classified as "bad". However, there is no human intervention so in terms of the Directive no action can be taken even though this should, by Directive definitions, be an area classified as "high" status. Similar results were obtained during an investigation near Oslo where the hydrochemical status was "good" following remedial work on a sewage treatment plant, but the natural fauna prevented the regrowth of macrophytes by excessive grazing – a natural condition – leading to a "poor" classification of biological status. What should the correct classification be in such cases? These questions will require answers quickly.

Groundwater objectives

Groundwater is treated in a different way when it comes to objectives. There are no ecological objectives. Instead a single objective is specified for quantitative status and for the quality of groundwater. Annex V (2) sets out the relevant criteria.

Good quantitative status is defined as the condition when *the available groundwater resource is not exceeded by the long-term annual average rate of abstraction.* This definition is qualified by the need to ensure that human intervention does not cause an adverse effect on surface waters, for example by reducing the base flow of rivers, nor must it affect terrestrial ecosystems which rely upon the groundwater in some way. Abstraction must not induce saline intrusion.

Good chemical status is slightly more complex. The chemical status is linked directly to quantitative status in that saline intrusion must be absent, and the chemical quality must not adversely affect any associated surface water. The more complex objective concerns the need to comply with the provisions of Article 17 (strategies to prevent and control pollution of groundwater). A particular issue is the requirement for the groundwater chemical quality to meet the specifications of any Community legislation which is applicable to groundwater. As, in general, groundwater which is used for drinking purposes will be categorised as a water requiring protection, the Community legislation applicable to protected areas will also provide the quality standards to be achieved under the definition of good chemical status. This will therefore be, as appropriate, the quality parameters of the Drinking Water Directive 80/778/EEC as amended by Directive 98/83/EC; the Directive on approximation of the laws relating to the exploitation and marketing of natural mineral waters (80/777/EEC) as amended by 96/70/EC; and the limits applicable to nitrate vulnerable zones under Directive 91/676/EEC. These Directives specify limits for approximately 49 chemical and physicochemical parameters and 5 microbiological organisms.

References

Chave P.A. (1993), Ecological aspects of water quality regulation in England and Wales. *Aquatic Ecosystem Health*, **2**, 65-71.

De Brabander K, Vanhooven G, Ringele A (1992), Newman P.J., Piavaux M.P., Sweeting R.A. (1992), ibid.

OECD (1982) *Eutrophication of Waters: Monitoring, Assessment and Control*, Organisation for Economic Cooperation and Development, Paris.

Pauw N.De, Ghetti P.F, Manzini P, Spaggiani R (1992), *River Water Quality – Ecological Assessment and Control*, Office for Official Publications of the EU, Luxembourg.

UNECE (1992) *Convention of the Protection and Use of Transboundary Watercourses and International Lakes, Helsinki 1992*, United Nations, Geneva, 1994.

WHO (1997) *Water Pollution Control*, E & FN Spon, London.

8

Programme of measures

INTRODUCTION

In order to achieve the water status that is identified as an environmental objective for each water body under Article 4, the Directive specifies that a *"programme of measures"* must be introduced. The measures are described in Article 11. They will form an integral part of the river basin management plan. The programme of measures represents a new framework for controlling activities within the river basin district, and Article 11 sets out a regime which includes the adoption of all appropriate measures for the control of pollutants and the prevention of water pollution, and for the use and reuse of water.

In devising the programme the data collected as a result of the survey of the river basin is used to identify the measures that are needed. Member States can devise each programme as a set of individual measures for each river basin district or they can choose to apply generalised legislative or other instruments to cover all river basins, including those parts of international basins which lie within their territory. The Directive specifies two types of measures, which it calls *"basic measures"* and *"supplementary measures"*. The programme is obliged to adopt basic measures and may choose to develop other supplementary measures.

BASIC MEASURES

Basic measures include the implementation of a number of directives that are already in force. These are set out in Annex VI and include the following:

Table 8.1 Directives which must be enacted under basic measures.

Directive	Designation
Bathing Water	76/160/EEC
Birds Directive	79/409/EEC
Drinking Water	98/83/EC
Major Accidents (Seveso)	96/82/EC
Environmental Impact Assessment	85/337/EEC
Sewage Sludge	86/278/EEC
Urban Waste-water Treatment	91/271/EEC
Plant Protection Products Directive	91/414/EEC
Nitrates	91/676/EEC
Habitats	92/43/EEC
Integrated Pollution Prevention Control	96/61/EC

These directives all have an impact upon and assist in the matter of water protection and several apply directly to water bodies. For example the Bathing Water Directive applies to discrete identified areas of water where bathing is practised. Individual bodies of water such as these will also be identified under the Framework Directive. The Birds Directive, Habitats Directive, Drinking Water Directive, and Nitrates Directive will be relevant to the issues relating to Protected Areas under this Directive.

The impact of introducing this requirement on Member states should be minimal, as all these directives should already be transposed and in the process of being implemented. There will be a need to identify their application within the river basin districts, however, and an administrative task to include and describe them within river basin management plans.

Combined approach

An innovation in water pollution control introduced by this Directive is the adoption of the concept of known as the "*combined approach*" for the control of polluting discharges.

Until recently there have been two quite separate views on the philosophy of pollution control in the aquatic environment. There has long been a view in many European and other countries that discharges into water should be controlled by deciding upon maximum limits of polluting substances which should be allowed to be discharged depending upon the industry involved and the constituents of the effluent. Limits are placed in the permit granted to the industrial plant. This approach is known as the emission limit value approach. The impact on the environment is not essential to the derivation of the limits, and the dilution or self-purification properties of the water are ignored. Rather, the value is derived by examining either what is achievable by the process or the toxicity of the substance in the receiving water. In some countries the use of best available technology to reduce concentrations of toxic materials in the effluent is required to be adopted by the plant before a permit is granted. In this case the emission limit value would reflect what could be achieved by the plant. This is the approach now endorsed in the Integrated Pollution Prevention Control Directive. In other cases the toxicity of the constituents and their effects on the water body is taken into account before deriving an appropriate limit. Limits may be set on a plant by plant basis or, more commonly, they are set on a national or even European basis, so that all industrial installations of a certain type discharge contaminants at the same concentration.

The alternate approach is known as the "*water quality objective*" approach. This system of control focuses on the receiving water body. Dependent upon the use of the watercourse, there is a maximum concentration of substances in the water which could be tolerated without being harmful to the aquatic environment. In other words, the water quality objective is a minimum water quality that will allow its use to continue unimpaired. In this situation a discharge of toxic or polluting substance is limited to the quantity that will allow the receiving water to remain within its quality objective concentration. The advantage of this approach is that both point sources and diffuse sources of pollution are taken into account, and the individual situation with regard to dilution and self-purification may be taken into account when the discharge limits are calculated, and it is not necessary to achieve the same level in each individual discharge. In both cases the discharge must be granted a permit before the plant may be allowed to make its discharge. Some of the early EC directives embodied a form of water quality objective approach (such as the Bathing Water Directive) but these were strongly use-related and set common limits on the water quality.

Now, there is realisation that both approaches have their merits, and indeed used together provide a stronger overall level of control than either individual system. In the EU Water Framework Directive Article 10 specifically requires the use of a combined approach as part of the programme of measures.

The approach is to be set up using as a basis the controls set out in a number of other directives including IPPC, Urban Waste-water Treatment and Nitrates, several directives relating to product controls (Plant Protection Products for example) and the set of directives which are daughter directives of Directive 76/464/EEC related to the discharge of dangerous substances into the aquatic environment. These are listed in Annex IX. This final set of directives already contains both emission limit values and quality objectives for the substances, but they are used as separate alternative ways for controlling pollutants rather than embodying an integrated approach.

When this approach is used, the most stringent values must apply in every case. Thus if a quality objective is specified in a particular legislative instrument, and this would lead to more stringent emission levels than would be acceptable under best available techniques or best environmental practice, then it is the quality objective which becomes the controlling influence, and the emission limits would have to be adjusted so that the quality objective is achieved.

The mechanisms for introducing the combined approach have to be implemented by 2012.

Implications of the combined approach

Through this Article a clear link is made between regulatory control of activities and the achievement of environmental objectives. Most countries have implemented a regulatory control mechanism for their industrial processes in order to reduce pollution, and many countries have adopted systems of best practice for such activities as agriculture. However, the direct link between these controls and the achievement of a quality of environment that sustains a good, and close to natural, biological abundance has not been made before.

Article 16 (priority substances) and Annex VIII extend the linkage of the EU Water Framework Directive with IPPC and pollution control policies. The list of main pollutants given in Annex VIII mirrors the group of substances for which community emission limit values will be set under Article 18 of the IPPC Directive and for which point source discharges will be controlled through IPPC authorisations. The EU Water Framework Directive does not, however, limit controls of these substances to the larger types of installation which will fall within IPPC. Article 11(3)(g) requires within the programme of measures, that controls on these substances by prior regulation, such as prohibitions, authorisations or the use of general binding rules, to be adopted for their control, and Article 11(3)l) specifies that measures must be taken to prevent losses from technical installations and to prevent accidental pollution incidents including the installation of early warning systems. Article 11(3)(k) aims to eliminate surface water pollution by the priority substances identified under Article 16 (see Chapter 12).

The implication of this set of implementation tasks is that the principles of IPPC will be extended by the EU Water Framework Directive to all installations, even those which are too small to be included in the full IPPC control procedure.

A discussion of the principles of IPPC and its implementation is beyond the scope of this book. However, most countries have transposed the Directive and are now in the midst of the implementation programme. Some countries have used the principles of IPPC in earlier legislation, so are familiar with the needs and difficulties of such a system. The UK has had a system known as Integrated Pollution Control (IPC) in place since the passage of the Environmental Protection Act in 1990. In France the Classified Installations Regulations have operated since 1976, and in Finland the Environmental Permit Procedure Act of 1991 operates a similar integrated approach. Most other countries have controls on installations that could be modified to meet the requirements of IPPC and thence towards the EU Water Framework Directive requirements.

Controls on diffuse sources of pollutants are more difficult to establish, but are necessary as many of the main list of pollutants and indeed some of the priority substances may also enter water bodies from diffuse sources. The programme of measures must address this problem. Countries vary in their adoption of measures to deal with diffuse pollution. In some cases diffuse pollution is controlled by regulation, for example by the use of prohibitions on the use of substances. In Sweden, for example, the Act Relating to Chemical Products regulates the use of chemicals by obliging those who handle import chemical products to take precautionary measures to prevent or counteract adverse effects on the environment. It also encourages the substitution of hazardous chemicals by less hazardous ones wherever possible. There are separate ordinances covering individual products. The Act for the Management of Agricultural Land also contains provisions aimed at environmental protection. Finland has a similar Chemicals Act (774/1989) controlling chemicals that are dangerous to health and the environment. Diffuse pollution is a difficult matter to resolve. The Plant Protection Products Directive, which has been implemented throughout the EU is one example of a general legislative mechanism for controlling substances that could cause diffuse pollution, by examining the overall likely impact of a particular substance on, in this case, groundwater and if the effects are too great, to ban its use.

In some countries the use of Code of Good Practice are given prominence. In the UK, the Code of Good Agricultural Practice stipulates the precautions that must be observed by farmers to prevent both point source and diffuse source pollution. Adherence to the Code should go a long way to avoid pollution, but if a problem occurs and a farmer is taken to court, the degree of attention he has paid to the Code is taken into account. This Code has been given statutory status. There are other relevant codes such as the Code for Forestry Activities and a Code on Pesticide Use, which can assist in reducing the incidence of diffuse pollution.

Waste management is an area mentioned as being important in the IPPC Directive, and it is of equal relevance to the water environment. Waste disposal options such as incineration and landfill both have potential impacts on water quality. There are EU legislative instruments covering both of these aspects, and a wider waste strategy which sets out rules for the management of domestic and hazardous wastes in the Waste Framework and Hazardous Wastes Directives (EU 1991a, 1991b). It is important to recognise the potential problems from the pollutants listed in the EU Water Framework Directive that occur in solid wastes and that may end up in surface or groundwaters. Adherance to the provisions of the Waste Sector Directives must also be seen as an essential part of the programme of measures. For the protection of groundwater, the provisions of the Landfill Directive are particularly important and it was concern over groundwater pollution that led to many of its requirements.

Recovery of costs

A further innovation to water management is the requirement to devise and adopt a cost recovery system to ensure that water pricing policies act as incentives towards efficient water usage so as to "*contribute to the environmental objectives of the Directive*" and to recover "*an adequate contribution*" to recovery of the costs of water services by the main user groups – industry, agriculture and households. The "polluter pays principle" must be applied. In order to implement this requirement the economic analysis undertaken as part of the survey of river basins conducted under Article 5 is used as the basic data set. The means of achieving this must be described in the river basin management plan.

Recognising that Member States have very different financial regimes at the present time, Article 9(4) rather surprisingly allows them to disregard this requirement in so far as they act "within established practices" and for a "given water-use activity" provided this does not compromise the purposes and achievement of the objectives of the Directive. If the Member State chooses not to adopt such financial measures the reasons must be reported in the river basin management plan. However, the use of other funding methods to achieve the objectives is permitted so that Member States are not inhibited in their financial methodologies to implement the Directive. Chapter 14 discusses economic options further.

Sustainable water use

Article 11(3)(c) specifies that measures to promote an efficient and sustainable water use must be introduced in order to safeguard environmental objectives. Although a "basic" measure such a requirement is very non-specific. Such measures could include the introduction of charging regimes under Article 9 (which requires Member States to invoke water pricing policies that provide adequate incentives to use water resources efficiently, but on the other hand, Member States are not obliged to introduce such specific schemes according to Article 9(4)). It could involve the voluntary introduction of water re-use policies such as that introduced in the UK under its Waste Minimisation initiatives (CEST 1995). Several countries have investigated groundwater resources and chosen to introduce licensing schemes for abstraction, with limitations on the amount that can be withdrawn with both financial and legal penalties for over-abstraction. The important issue for this measure is the need to assess what is sustainable by reference to the resource availability and the effect that over-use or non-sustainable abstraction may have on the waters.

Drinking water

In the programme of measures, drinking water resources are especially identified for protection. Article 7 of the Directive requires the identification of all bodies of water, surface and groundwater, that are, or may in the future, be used as a source of drinking water for 50 persons or more, or where the rate of abstraction is more that 10 cubic m per day. Deterioration in the quality of these water bodies must be avoided so that less treatment is required to render the water suitable to drink. The treated water must also meet the standards in the Drinking Water Directive (80/778/EEC as amended by 98/83/EC). The specific measures needed for protecting these waters are not described, but the Directive indicates that "safeguard zones" may be established. Waters that are used for drinking water sources are already well documented in most countries of the EU and elsewhere in the world. It is common to apply special protective measures for such water sources. These range from general provisions in legislation to the use of various types of protective zone. Such zones are widely used for the protection of groundwater sources and in some countries "perimeters" have been set up to protect surface waters.

The implications of this requirement for Member States are thus that each individual source which is actually, or possibly, utilised for drinking water must be identified, catalogued and surveyed to determine what protection is needed. The identification of these waters should be accomplished within the survey undertaken for Article 5, because part of the survey is connected with human activity in the catchment area, and with an analysis of water use. Once identified, the necessity of protection must be assessed.

The establishment of protection zones for both groundwater, where they are commonly in use (see Chapter 10), and for surface waters, where they are much rarer, is one option. Under the Abstraction Directive (75/440/EEC) surface waters used for the abstraction of drinking water should by now have been identified in all the Member States, and measures introduced to improve water quality to at least Class A2, so that conventional water treatment processes may be used to bring the water up to the standards required by the Drinking Water Directive. Measures to safeguard such quality should have been introduced. These may be the increased use of existing legislative controls on discharges, or activities in the catchment zones of the surface waters, or it may involve special measures such as the creation of particular protection zones. France and Spain both have provisions in their national laws for the creation of zones (in France known as "zones de saufeguarde") and these are well used. In the UK there is provision in the Water Resources Act of 1991 to set up protection zones but to date only one such statutory zone has been approved – on the River Dee, which is heavily used for water supply and is subject to extremely damaging pollution events from industry. The problem with establishing such zones is that of resistance from industry and others who have to take special precautions to guard against pollution incidents, at some cost to themselves. In the longer term, the use of land-use planning legislation is available as a means of preventing the development of possibly hazardous installations in such zones, or more generally, provided the catchment areas of the water supplies are known to the planners, and the risks are clearly understood. Some countries elsewhere in the world have set up development corridors that specifically prohibit certain types of development to safeguard water supplies (for example, Western Australia). Countries in mainland Europe rely on rivers that cross country boundaries for their surface water supplies. In these cases the establishment of special protection measures is very difficult and relies heavily on inter-country cooperation. This is partly the reason behind many of the conventions now in place. The Framework Directive will place even greater emphasis on the need for such cooperation in the case of water-supply rivers and the less well developed issue of groundwater protection.

Water abstraction and recharge

An important matter for the programme relates to the amount of water that is withdrawn or put back into the river basin. Member States are obliged to identify within the surveys carried out for Article 5 the impact of water use including an economic analysis. As a result of this, the water use patterns should be established, and water abstraction and return discharge points identified. All impoundments for whatever purpose should be identified, as their resultant lake or ponds must be assessed for their possible inclusion in the list of water bodies. The Directive requires a register to be set up of all these abstractions and impoundments and the introduction of a prior authorisation procedure for their construction or operation. Most countries already have such systems in use as a means of protecting their water resources, and generally subject to a charge or water tax. For example in the UK any person wishing to abstract water for public supply, industrial use or irrigation must apply to the Environment Agency for an abstraction licence. There is a charge made depending upon the amount of water abstracted and its use. The Agency is able to place conditions on the abstraction as to the quantity and purpose, and the proposal is open to public consultation before being granted. Once granted, the licence details are placed on a public register, and the abstractor has to provide regular details of the quantity of water taken. This enables the Agency to maintain control over the quantity of water taken from a resource, and to protect its long-term viability using both the regulatory controls over quantity and the financial controls inherent in a charging regime. There are exemptions for very small individual wells, but even these are subject to registration. Artificial recharge activities must also be controlled under this Directive, but again, as these are generally large-scale operations carried out by water authorities and subjected to licensing procedures, there is unlikely to be a significant impact in this requirement. In Germany the Federal Water Act has as its objectives the maintenance of water bodies and their technical development to safeguard their aesthetic and ecological functions. Water flow characteristics are determined by the physical state of the water course and the aim of the legislation is to maintain them in such a state as to secure adequate water flow.

Groundwater

Measures to protect groundwater are also required, and are specifically mentioned in Article 11(3)(j). Chapter 10 discusses this in some detail. The Article prohibits the direct discharge of pollutants into groundwater, but it permits prior authorisation of a number of specific activities related to the re-injection of waters that have been extracted for particular purposes such as dewatering for mining or construction, exploration for oils and injection for the storage of gas. Such discharges are only allowed if the groundwater is unsuitable for any other use. However. the injection of small quantities of substances for characterisation, protection or remediation of groundwater bodies is permitted. Construction and civil engineering works which come into contact with, and could therefore influence, the water table require authorisation and general binding rules, have to be drawn up by each Member State for such activities. This could become a significant administrative and possibly technical burden in counties where there are high water tables and usable groundwater is widespread. The programme of measures will have to indicate the content of such rules and set up as procedure for their operation.

Floods and droughts

In Hungary in March 2001 floods occurred in the River Tisza floodplain that overwhelmed defences that had been developed over a period of over 100 years and caused widespread damage. In the UK in the autumn of 2000 large areas of the country suffered much more flooding than has been known for many years due to long periods of excessive rainfall. Were these catastrophies a result of climate change or from lack of investment in flood defence works or poor land use planning decisions? It is not possible at this stage to give a precise answer. Nevertheless, in terms of river basin management, the possibility of flooding must be considered a relevant feature of the river basin plan. One of the purposes of the Directive, as stated in Article 1(e), is to contribute to mitigating the effects of floods and droughts as a means of contributing to the provision of good quality water, reducing pollution of groundwater and protecting territorial and marine wasters. The effects of floods and droughts on water quality and status are important as they are extreme events in terms of their effects on the ecology. The programmes of measures should take account of the needs for measures to mitigate such problems.

SUPPLEMENTARY MEASURES

In addition to the above basic measures, the Directive permits the adoption of supplementary measures as a part of the programme. Possible measures are outlined in Annex VI and these are listed in Table 8.2.

Table 8.2 Supplementary measures.

Legislative instruments
Administrative instruments
Economic or fiscal instruments
Negotiated environmental agreements
Emission controls
Codes of good practice
Re-creation and restoration of wetlands
Abstraction controls
Demand management measures
Efficiency and reuse measures
Construction projects
Desalination plants
Rehabilitation projects
Artificial recharge of aquifers
Educational projects
Research, development and demonstration projects
Other relevant measures

Many of these supplementary measures are fundamental components of the basic measures. For example, legislative and administrative instruments will be a part of the legal framework for most of the basic measures. Certain emission controls, abstraction controls and codes of good practice are dealt with under basic measures but there may be other measures of this type unique to individual country situations and Article 11(4) specifically advises that these should be used where appropriate. Supplementary measures permit other aspects of control beyond those generally associated with Community legislation to be introduced.

ENFORCEMENT

An important issue, essential to the success of any programme of measures, and which is emphasised by Article 23, is the need to ensure that measures are implemented and obeyed. In order to achieve this Article 23 requires Member States to determine penalties applicable to breaches in the national provisions adopted for the purposes of the Directive. The penalties must be effective, proportionate and dissuasive. All countries in the EU set penalties within their legal systems for breaking the law. In the environmental protection field penalties vary widely, depending upon the precise nature of the offence. For example in the UK the penalty for causing pollution under its Water Resources Act ranges from a maximum fine of £20,000 in a lower court to an unlimited fine and up to two years imprisonment if the offence is taken to a higher court.

A fine of £1,000,000 has already been exacted in one case. In other countries breaches of permit conditions may be considered an administrative matter. In Denmark the Environmental Protection Law involves fines and up to two years in prison, whereas in Germany it is limited to a fines procedure. In the Netherlands the environmental legislation is supported by an Economic Offences Act the maximum penalties of which are heavy fines and up to six years imprisonment.

The enforcement agency must be clearly identified. In most countries the enforcement agency is the relevant environmental protection agency. Where enforcement action is required the agency usually carries out the inspection and sampling duties, and often, if the breach of conditions or the pollution offence is of a minor nature, action against the polluter is taken by the agency. Where the decision is taken to involve court action, some countries involve the police or the public prosecutor. In England and Wales all enforcement activities remain with the Environment Agency, which retains it own team of prosecution lawyers. At the other extreme, Italy has set up a separate operational Ecology Unit of the Carainieri which is placed at the disposition of the Ministry of the Environment to undertake environmental enforcement (IMPEL 1996). Table 8.3 includes the basic issues that need to be addressed in setting up the enforcement regime, whilst Table 8.4 shows possible responses. One of the key points in the process is the promotion of compliance. The EU Water Framework Directive uses the terms effective and dissuasive. In Table 8.3 compliance promotion is shown as a positive activity.

Table 8.3 Initialising an enforcement regime.

Policy Planning	Identify and analyse environmental issues.
	Set priorities.
	Establish resources
Legislation	Pass laws
Permitting	Ensure that enforceable permit conditions are set
Compliance Control	Establish what is important for environment,
	Check compliance
	Use protocols, on-site visits, code for inspectors
	Report consistently
Compliance promotion	To achieve compliance in cases of non-compliance, convince operator to conform rather than use legal enforcement mechanism
	Consequences of non-compliance must be serious
	Allow a transition period

Table 8.4 Enforcement response

Enforcement response	Court proceedings
	Temporary shutdowns or other penalties
	Avoid negotiations
	Ensure the verdict will be in the regulators favour to avoid loss of credibility
	Violations of self-monitored limits should require pre-established corrective measures
Assessment and Feedback	Adjust regulations if necessary in light of experience
Integrated permitting	One permit to cover all environmental issues
Compliance verification	Use integrated permits to reduce regulator's workload
	Prescribe self-monitoring and record keeping
	Automatic measurements and reporting
	Data should serve clear objectives for regulators and operators

IDENTIFICATION OF CRITICAL POINTS IN THE PROGRAMMES

The survey undertaken for Article 5 will identify the details of the problems in the river basins. The programme of measures will deal with them but it is helpful to know the particular issues that could lead to failure and to be able to verify that these are being dealt with. As a means of identifying measures to safeguard water supplies, the World Health Organisation is suggesting the use of a technique first used in the food industry to identify critical issues. Known as Critical Control Points they are part of the HACCP approach (Hazard Analysis Critical Control Point) developed by the food industry in the 1960's (Godfrey 2000). An analogous approach in drinking water management is currently being discussed, particularly by the "Microbiology Working Group" for the revision of the WHO Guidelines for Drinking Water Quality. Introducing this concept into water management is attractive because it directs attention to the details of impacting issues and good management practices to preventing failure. End-product testing for compliance to guideline values then has a new function – the emphasis is on testing performance of the system. This may be particularly useful in settings with restricted access to analytical methods particularly in the context of biological assessment: "next best guess" or "default values" may be used where analytical options are lacking, e.g. critical site inspection with a focus on performance indicators. The assessment of the system would then not be *"we don't know because we can't measure"*, but rather "*though analytical data are lacking, structures or practices observed (e.g. application of liquid manure on frozen ground) indicate a risk of water contamination (e.g. with pathogens and nitrate)*". The estimate can be improved by knowledge of hydrology and the type of soil.

Within each of the measures in the programme steps for attaining good water quality may be developed and within each of these steps, management measures may be designed as barriers against impacting agents for example nitrate management in agriculture as the first step of groundwater protection. One or more control points ("CPs") may be identified for a given management measure, i.e. for nitrate management in agriculture a CP might be the existence of a farm nutrient management plan, or regulations for stock density, or for manure application. These general CPs may be further specified as appropriate for a given setting, for example specific seasonal restrictions of the application of manure.

Following the general principles of multiple safety barriers, there should not only be one control point that is critical for the whole process but rather a sequence of control points. The difference between the concept of a "barrier", or "management measure" and that of a "control point" is that there is opportunity to verify the effect of a control point. If management plans identify CPs in conjunction with the options for their respective verification, this may be a breakthrough in their implementation.

Thus, the development of a land use plan is not an example of a CP, but within this management measure, definition of protection zones would be a CCP, as would implementation of the land use plan. A CP in this understanding is a specific behaviour of physical measure that can be put in place and verified in the larger context of management measures.

CPs may also be very useful for institutional management issues, because of the introduction of verification for these as well, e.g. a CP of "*incident management plan*" and verification through "*documentation of annual incident management exercises*", or a CP for the "*capacity to undertake actions*". Table 8.5 shows an array of very different examples for CPs developed by WHO (Chorus 2002) in protecting groundwater for health and options for verification.

Table 8.1 Examples of the use of critical control points for protecting groundwater

Management issue	Examples of Control Points	Examples of options for their respective verification
Avoiding disease agents from any human activity	Drinking-water protection zone	Periodic site inspection
Avoiding disease agents from any human activity	Reliable groundwater moni-toring data for disease agents	Records on straff training; certification of good laboratory practice (GLP)
Avoiding disease agents from wastes	Waste classification and separation	Docket and sample analysis of waste truck content
Avoiding disease agents from wastes	Sealing of the dump	Upstream and downstream contaminant analysis in groundwater samples
Avoiding hazardous chemicals	Effective containment during transport and storage	Regular inspection of containments
Avoiding hazardous chemicals	Effective emergency plans for spills	Records of staff training for emergency intervention
Avoiding disease agents from human excreta	Distance between latrines and wells	Regular site inspection and/or upstream and downstream analysis of groundwater (GW) samples
Avoiding disease agents from human excreta	Low leakage from sewers	Regular inspection of sewer condition
Avoiding disease agents from agriculture	Stock density adequate for local soil and hydrological conditions	Control of farm book-keeping; site inspection
Avoiding disease agents from agriculture	Mineral fertilizer application adequate for local soil and hydrological conditions	Control of farm nutrient management plan; soil analysis

An essential element of the CCP concept is their specificity for different technologies and situations. The option of monitoring for verification on different levels of sophistication, according to the levels of available equipment, is attractive. It is important to establish who will carry out the verification (institutional issues are involved here) and review of the CPs, as this will in many cases need to be an intersectoral exercise, and community participation in verification is critical. Hazards will change, and new means to abate them are also likely to evolve. Therefore, the process of identifying CPs must be revisited and revised regularly.

SUMMARY OF PROGRAMME OF MEASURES

Figure 8.1 sets out pictorially the procedures within the programme of measures.

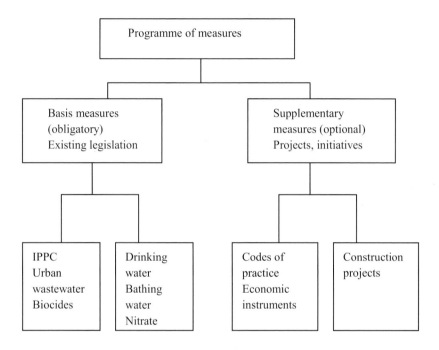

Figure 8.1 Programme of measures.

REFERENCES

CEST (1995) *Waste Minimisation*, Centre for Exploitation of Science and Technology, London.

Chorus, I. (ed) (2002) *Protecting Groundwater for Health*, WHO, Geneva, in press.

EU (1991a) Waste Framework Directive 75/442/EEC amended by 01/156/EEC and 91/692/EEC.

EU (1991b) Hazardous Waste Directive 91/689/EEC amended by 94/31/EEC.

Godfrey, A., (2000) Assuring microbiologically safe drinking water, AWWA Water Technology Conference.

IMPEL (1996) *Inspection and enforcement of environmental legislation in some EU countries and regions – a systematic descripton*, Danish Environmental Protection Agency, Copenhagen.

9

River basin plans

PREPARATION OF RIVER BASIN PLANS

One of the potential strengths of the approach outlined in the Directive is the move towards forward planning for water protection. Whilst there are many ways of regulating activities that may cause a deterioration in water quality or quantity, these are often used in isolation from the overall position with regard to the environment, and tend to deal with individual issues. For example, discharges of polluting materials are regulated either to ensure that particular water quality objectives are met in a specific stretch of the river, or because a particular industry is seen to cause problems and limit values are prescribed for that industry nationally or internationally. Agricultural pollution is often dealt with on a farm by farm basis and individual solutions to local problems found – perhaps agreeing a farm management plan or, for broader issues, by banning a particular pesticide or issuing best practice guidance often without reference to the wider picture. It is relatively rare for land use planning to be seen as a fundamental issue of water protection, and there are many examples of inappropriate developments leading to derogation in water quality and water quantity.

Under the terms of the EU Water Framework Directive this situation should change. The competent authorities are obliged to prepare a river basin management plan for each of the river basin districts that lie within a Member State's territory. As outlined earlier, a river basin district may comprise one river or the catchment areas of more than one river. It may apply to cross-border rivers with parts of a district in another country. Because of this difficulties may arise in deciding how extensive the river basin plan should be. The Directive recognises these problems and whilst an overall catchment plan must apply to each river basin district – perhaps thought of as a "master plan" for the district, although the Directive does not acknowledge it as such – the plan may include more detailed plans for sub-catchments, or even individual rivers, where these are small and have been combined to build up the river basin district. Such subsidiary plans can also cover diverse issues, particular problems, sectors or water types within the river basin. The Directive specifies that only one plan should be prepared for each river basin district, however.

Where rivers cross international borders the situation is more complex. If the river falls totally within the territories of the EU, the Directive specifies that the countries concerned *shall* cooperate in producing a single management plan for the whole river, but it also indicates that if this is not produced, plans should be produced for those parts of rivers that fall within a State's territory. This is a slightly confusing legal requirement, the word *shall* indicating a mandatory requirement but the Directive takes account of the fact that Member States are sovereign states, and may not be able to collaborate to the degree required to produce a single plan. Again, if the river basin extends beyond the boundaries of the EU, Member States must at least produce a plan covering those parts of the river basin district falling within their territories, but they should endeavour to collaborate with other States in producing a single plan covering the whole river.

River basin plans have to be published within 9 years of the coming into force of the Directive, with a review every 6 years after this time. However, there is a significant element of public consultation built into the preparation process that reduces the time available for their production.

CONTENT OF A PLAN

A river basin plan has to cover a minimum set of issues. The details are set out as Annex VII. Many of these issues are dealt with as a result of implementing other Articles of the Directive so the river basin management plan is seen as an integrating step in the framework, pulling together much of the implementation activity. Table 9.1 shows a summary of the plan contents.

Table 9.1 Summary of river basin plan contents

Contents
General description of river basin district (Art 5).
For surface waters: maps of location and boundaries; eco-regions and surface water types, reference conditions
For groundwaters: maps of location and boundaries.
Summary of significant pressures and impact of human activity:
Point sources of pollution; diffuse source of pollution; pressures on quantity; analysis of anthropogenic impact on water status.
Identification of protected areas (Art 6)
Maps of monitoring networks and the results of monitoring:
surface water ecology and chemistry;
groundwater ecology and chemistry;
protected areas.
Environmental objectives (Art 4) including extensions and derogations
Summary of economic analysis of water use (Art 5)
Summary of programme of measures (Art 11)
Summary of measures to implement EU legislation for water protection
Details of practical steps to recover costs of water use (Art 9)
Measures taken to protect drinking water sources
Summary of controls on abstraction, impoundment, point source discharges and other activities (Art 11)
Authorisations of direct discharges to groundwater (Art 11)
Summary of measures to deal with priority substances (Art 16)
Summary of measures taken to prevent accidental pollution
Details of supplementary measures needed to meet environmental objectives
Measures taken to reduce marine pollution (Art 11(6))
Register of more detailed plans for sub-basins, sectors, issues or water types
Summary of public information and consultations and their results
A list of competent authorities
Contact points for obtaining more information under Art 14(1), control measures described in Art 11(3) and actual monitoring data

These items form the basis of the first river basin management plan. At subsequent reviews some further information is required concerning the progress of the plan. These are:

- a summary of any changes to the plan itself;
- a report on the progress towards the achievement of environmental objectives, including a map showing summary monitoring data and an explanation as to why particular objectives may not have been achieved;
- a summary of measures that have not been undertaken; and
- a summary of any interim measures adopted since the plan was published.

EXISTING PLANNING EXPERIENCE

What experience is available to date in such a task? A number of countries have developed techniques for catchment management planning over the past few years. Some of these examples match closely the proposals in the new Directive. There are examples of existing river basin plans that apply solely to rivers within a country's territory, and several examples of plans that rely upon cooperation between countries in dealing with cross border rivers. A number of agreements and action programmes exist that cover the maritime situation, but these do not generally relate to a river basin. A small number of examples may be found of plans to manage lakes, including a few lakes that cross international borders. Again these are not usually related to the whole river basin.

Within country river basin plans

United Kingdom

In the UK the water industry has, since the early 1970s, been organised on a river catchment boundary basis, so that planning work has largely related to river basins. In the 1973 Water Act a system of integrated water management had been introduced under which the protection of water quality and quantity, and the management of the water supply and sewage treatment infrastructure had been the responsibility of public bodies known as river authorities, which had comprehensive control of all aspects of the water cycle. This included water supply and sewage treatment, pollution control, land drainage and flood prevention. Within the legislation there was a requirement for a comprehensive survey to be carried out to determine the needs of river catchments in respect of water resources, including quality and quantity assessments and the requirements for future infrastructure. This concept was seen by several other nations as a positive contribution to water management and it was emulated in a number of places.

In 1989 the water industry in England and Wales was reorganised in such a way that the 10 regional water authorities were amalgamated into a new regulatory body known as the National Rivers Authority (NRA), which retained the previous river basin boundaries. Water supply and sewerage services were devolved into a number of private companies. Whilst the NRA had no formal duty to prepare plans based on river basins, it implemented a system of catchment management plans for the rivers under its control. Catchment plans were prepared by carrying out a survey of all the activities in the river basin, and relating these to the perceived problems of water quality and quantity, consulting with local organisations that might be affected by any proposals for improving the water environment and producing an agreed plan of action for the river basin. All of the major rivers received this attention over a period of years so that each major river basin had a published plan and timescales for implementation.

A further amalgamation of organisations took place in 1996 as result of the 1995 Environment Act, bringing together the regulatory bodies concerned with water, waste and industrial installations and the Environment Agency. The catchment planning process has been extended to a more detailed level, to include all of the new Agency's functions. Still based on river catchments, these cover the rivers or the sub-catchments of larger rivers and plans are produced that are known as Local Environment Agency Plans (LEAPs) in a process that includes considerations of all activities within the river basin or sub-basin. The LEAPs are non-statutory documents that describe the river basins and identify problems and their possible solutions (Taw 2000). The plans cover a period of 5 years. The implementation of each plan is monitored and a report on progress is published annually. The annual review identifies any additional actions needed and amends those that are no longer required. A major review of the plan takes place every 5 years. The format of these plans follow the outline river basin management plans of the new Directive fairly closely. They begin with a description of the LEAP area, which is derived from a survey of the river basin and existing data sources. The description includes an outline of the physical features such as the hydromorphology, geology and hydrogeology including the identification of the main aquifers, details of soil types and river flows, information on wildlife habitats, and heritage issues and a description of land use and activities in the catchment. Targets for water quality are determined using the Environment Agency's river quality objective scheme. This uses a five level hierarchical class system ranging from very good quality to poor quality using, primarily, chemical parameters (see Chapter 7). It is a principle of the management regime that the river quality objectives must be achievable and sustainable, and it must be possible to identify what needs to be done to reach and maintain the quality objectives in the future. Complementary to the formal chemical based river quality objectives is a scheme for monitoring the biological quality of the river using macro-invertebrates as the indicator organism. The issues affecting the river quality and the environment in general are identified and policies for their remediation are devised. The implementation of such policies depends upon dialogue and cooperation with those organisations that cause or control the problem activities.

Germany

In Germany there is already a requirement within the Federal Water Management Act of 1957 for the *Lander* to draw up water management framework plans for river basins and conurbations. Guidelines have been issued that lay down the format for these (Regulation 1984). The plans must include information concerning water availability; water demand; water balances between availability and demand; principles to be adopted for pollution control; and flood protection requirements. The plans require an assessment of the river basin in much the same way as is required by the new Directive. The natural characteristics of the basin and human activities are surveyed as a basis for identifying the availability of water and for calculating present and future water needs. The plans identify present and future management needs for flood and resource protection. In addition to the overall basin management plans, *Lander* must also produce more specific plans for the management of surface water bodies.

A 1978 addition to the Federal Water Management Act requires the basin plans to concentrate on water resource protection, but those plans that are developed for individual water bodies (*Bewirtschaftungsplane*) concentrate on water quality. They are produced in situations where future drinking water supplies are at risk or where planning is needed to meet international obligations. The content of such detailed plans include target ecological quality values, indications of the use to which the water is to be put, quality objectives and parametric values that must be met and the measures that must be taken to achieve these objectives, timescales for the plan to be completed and other management issues that may be required to protect the resources. Such plans utilise a water quality objective approach to the determination of emission controls.

In addition to the river basin plans, separate plans have also to be drawn up concerning effluent disposal, and in some of the eastern *Lander* plans for redevelopment of water supply systems are necessary. These plans include many of the issues that must be examined in the survey of human impact, and economic review required under Article 5 of the Framework Directive such as the description of wastewater and water treatment plants, pollution loads in the river system, and water bodies that are used for drinking purposes, and the identification of areas where additional protection is needed in the river basin. Amalgamating such plans into a composite river basin district could form the basis of the planning process for the new Directive.

Portugal

Portugal has two levels of river basin planning – a national plan and individual river basin plans (Correia *et al.* 1998). The responsibility for their preparation lies with an organisation known as the National Institute for Water (INAG). There are 15 river basin plans for the river basins of the larger rivers in the country. The river basin plans cover a period of 8 years with a review period of 6 years. The national water plan covers a longer period, 10 years with a review in 8 years. This is not too different from the proposals in the framework Directive. The plans are derived from a national law (decree-law no. 45/94) so that river basin plans already have a legal status. The law requires that planning should take account of the environmental, technical, economic and institutional aspects and aim at the optional use of resources to meet all needs. The issue of land use planning is emphasised with the requirement that this should be integrated with other issues such as environmental protection. The participation of the public and affected parties is also encouraged. These basic premises of the law fit well with the aims of the new Directive. Further, the river basin plans include a series of steps similar to the DPSIR framework outlined later in this chapter and which match the implementation steps outlined in the Directive. They include a survey of the river basin, the formulation of objectives (short, medium and long term), the development of an action framework with alternatives and priorities, and an implementation stage. The plan must contain the following items:

- A classification of watercourses and water bodies in terms of their use;
- Means to protect and enhance the river network;
- Means to protect and enhance groundwater resources;
- Application of the polluter pays principle by way of a regulatory tax;
- Definition of areas subject to land use restrictions;
- Identification of protection zones;
- Identification and prioritisation of water and wastewater structures;
- Actions needed to prevent floods and regulate rivers;
- Information on sediment movement.

The national water plan takes a countrywide view of the individual river basin plans. At this level, the national plan requires the integration of river objectives with national economic and social policies, and must include the means by which the plans of the individual river basins can be coordinated. At the national level, there must be a definition of priority areas, and where plans put forward proposals such as inter-river transfers to improve water availability, these must also be included in the national plan. Where rivers cross the international border with Spain it is the national water plan which is the tool for proposing and implementing the necessary coordinating measures.

Ireland

The Irish government has being developing river basin management concepts since 1997, and an example of one of a series of catchment based water quality monitoring and management systems being introduced is the 'Three Rivers Project' covering the rivers Boyne, Liffey and Suir (IWA 2001).

The establishment of such multi-sectoral, basin-wide and community-based systems is a response to historically disjointed, legalistic and non-participative approaches to water resource management and they are an attempt to transcend the restrictions of traditional local authority administrative boundaries. The pilot project which is the first to focus on river water quality embodies the concepts and objectives contained in the EU Water Framework Directive. Ireland, in common with many EU countries, has failed to halt decades of increasing levels of eutrophication of surface waters caused principally by phosphorous loading, and Irish authorities, concerned about river water quality, are now turning to catchment based strategies in an attempt to reverse the continually deteriorating trend in surface water quality. The 'Three Rivers Project' is promoting the benefits of an integrated and cooperative approach to the management of three important river systems in Ireland. The success of the 3 year project which commenced in September 1998 with a budget in the region of IR£3.5m, hinges on a number of key elements which include:

- the creation of an effective management structure;
- the establishment of comprehensive baseline information;
- the development of a competent public awareness campaign;
- the setting up of an integrated and targeted monitoring network including the incorporation of existing systems where appropriate;
- generating a specified and purpose-built Geographical Information System;
- estimating sectoral pollution loadings;
- implementing pollution abatement strategies
- auditing progress.

The project objective is to protect and improve water quality to conform with 'good ecological status'.

Project experiences and findings to date are interesting. For example, key data sets relating to hydrometrics, soils and fisheries were effectively unavailable. In addition, on-site automated sampling has proven to be very cost effective and appears to be generally essential if reliable pollution loadings are to be assigned; the automatic samplers that provide for both time composite and flow composite sampling captured 'event' results that could not have been achieved by grab sampling response and reported significantly higher total phosphorous loads in response to increased rainfall/flow than would have been predicted from the grab sampling records. The automatic sampling also highlighted detrimentally low oxygen levels during the hours of darkness for salmonids in eutrophic waters. An interesting point relating to agricultural best management practices is that up to 90% of pollution loading may be sourced from under 10% of a holding and, accordingly, proper risk assessment must be undertaken on a 'plot by plot' basis i.e. at micro level. The implications of the Project findings for agricultural, municipal and industrial policy are grave and one of the greatest challenges now is to organise and fund Irish river basin management systems as envisaged by the Framework Directive and to continue and build on the work which the 'Three Rivers Project' has achieved. National, regional and local authorities and interest groups have now recognised the benefits of an integrated and partnership approach at catchment level (being the natural and preferred unit of management focus) and are framing catchment based strategies for development and implementation into the future in recognition of the mandatory requirements of the EU Water Framework Directive. Approximately, seven 4-year River basin district projects are now under active procurement to take the Directive forward in Ireland under the guidance of project consultants, a steering group and a representative body of interested parties for each identified river basin district.

INTERNATIONAL RIVER BASIN PLANS

There are several examples of river basin plans that have been derived for cross border river systems. In Europe, the Rhine and the Danube, for example, have water quality plans developed over a number of years.

River Danube Environmental Programme

Countries riparian to the Danube had mostly developed local plans for their the reaches of the river passing through their own territories, but in 1991 at a meeting in Sofia, Bulgaria, these countries agreed to develop a series of studies and actions towards the establishment of an international agreement. This *Environmental Programme for the Danube River Basin* included reviews of activities and conditions within states, basin-wide studies of point and non-point sources of pollution and biological status, and detailed studies of individual tributaries. The conclusion in 1994 of the *Convention on Cooperation for the Protection and Sustainable Use of the River Danube* now provides a legal basis for protection of water resources. The signatories have agreed on the conservation, improvement and rational use of surface and groundwaters in the catchment area, to control hazards originating from accidents involving substances hazardous to water, flood and ice-hazards; and to contribute to reducing the pollution loads of the Black Sea from sources in the catchment area. Cooperation is agreed on using all appropriate legal, administrative and technical means to at least maintain and improve the current environmental and water quality conditions of the Danube River and of the waters in its catchment area and to prevent and reduce as far as possible adverse impacts and changes occurring or likely to be caused. An international commission has been set up to provide the framework for cooperation between the signatory States. The strategic action plan drawn up by a task force specifically charged with that duty covers the period 1995–2005. (Danube 1994). The countries have already agreed to use a number of the principles embodied in the Framework Directive, namely the adoption of the precautionary principle; the use of best available techniques to reduce pollution; the control of pollution at source; and the polluter pays principle, together with a commitment to share information and cooperate regionally in implementing the plan. The action plan has four goals: to reduce negative impacts of activities in the Danube river basin on the ecology of the river and the Black Sea; to maintain and improve water quality and availability in the river basin; to control hazards from spillages which occur accidentally; and to develop international cooperation in water management.

Comparing this plan with the principles set out in the Framework Directive, there is a close agreement. The plan covers the whole of the Danube river basin and includes river water, coastal water and also underground water; it refers to quantity as well as quality; it relies upon an integrated water management regime; and it identifies and sets water quality objectives for water uses. The information gathered to enable the plan to be drawn up forsees a number of problems associated with ecosystem health and water use. In the context of the Directive, water uses and the economic consequences of an inability to maintain those uses gives a good approximation of the economic study required by the Directive. The plan itself picks out the main problems as being microbiological contamination, which renders the water generally unfit for use as water supply for drinking, industry, irrigation or recreational use; high nutrient levels affect its use for drinking, fisheries, industry and recreational opportunities; the presence of hazardous substances affect all possible uses; and competition for available water also causes loss of habitat, and reduced supplies.

River Rhine Action Programme

The Rhine Action Programme is another European example of an internationally agreed river basin plan. This programme was set up after the disastrous Sandoz chemical pollution incident which occurred in 1986, with a view to increasing the controls on discharges to the Rhine in order to accelerate the rate at which these were reduced, and in order to improve water quality. The Action Programme had four main aims – to improve the quality so that indigenous fish could survive; to ensure that the Rhine could continue to be used as a source of drinking water; to ensure that sediments were not polluted by harmful materials; and to assist in the recovery of the North Sea. The Programme introduced a number of measures including a planned reduction in load by 50% between the years 1985 and 1995, action to prevent pollution incidents and a range of measure for the restoration of the Rhine ecosystem. The implementation of such a plan requires collaboration of all the signatory countries to the Agreement on the International Commission for the Protection of the Rhine against Pollution of 29 April 1963. The countries involved include Germany, Netherlands, Luxembourg, France and Switzerland.

River basin planning in the Netherlands

The Netherlands represents a country for which water protection planning is a very important issue and most of its river basin plans rely upon collaboration with other countries. Its main river network is at the lower extremity of large international rivers – the Rhine and the Meuse, the Scheldt and the Ems. Much of the country lies below sea level, protected by dyke sea defences, and much of the countryside is man-made – land that was released as a result of the damming of rivers and draining of wetlands. There are occasional massive flooding problems – for example in 1995 around a quarter of million inhabitants were forced from their homes due to flooding of the Rhine and Meuse. Water quality is a problem as agriculture is probably amongst the most intensive in the world, and is a cause of eutrophication of surface waters. As a result of these pressures, water planning is very important. In 1985 the Ministry of Water Management adopted the principles of an integrated water management scheme is in force through the adoption of general framework legislation which uses a technique known as the "water system approach". The "system" covers ground and surface waters, their direct surroundings and their flora and fauna. It aims to integrate ground and surface waters, their quality and quantity, and their usage. Water planning covers the quantity and quality aspects of both surface and ground water and there is extensive coordination with land use planning and other environmental and nature management plans. The Water Management Act of 1989 integrates the plans that are drawn up at three administrative levels – state, province, and water board – such that each level has responsibilities covering all of these functions, and each level has a management plan to deal with its responsibilities. The plans are not, however, based solely on a river basin district as is required by the Directive.

The country is so vulnerable to developments taking place on its main rivers in upstream countries that it was an early proponent of an international approach to water management. In addition to the Rhine Commission, it is also a signatory to a 1995 treatise with France, two Belgian regions and the Brussels region covering the rivers Scheldt and Meuse under which cooperation on the preparation of an action programme for the rehabilitation and improvement of these rivers will take place. It is also a signatory to an international plan involving Germany, Belgium, Luxembourg and France to tackle the problems of the high water levels in the Rhine and Meuse which requires that integrated measures must be taken involving both water management regimes (such as the building of flood prevention structures and flow regulation), and land use planning (involving measures such as agriculture, urbanisation and forestry).

Other examples of international river basin management programmes

The river basins of the Moselle and Saar, which are shared between Germany and France, are managed through international commissions and various protocols. Two International Commissions for the Protection of the Saar Against Pollution, and the Protection of the Moselle From Pollution, were established in 1961, and additional cooperation, particularly in respect of flood control, was concluded with Luxembourg, France and Germany in 1984 and 1987. There are similar commissions covering the Elbe (1990) and the Oder. These latter two involved the European Community as an international body, and also took up the issue of river basin planning with States that are currently outside the EU – Poland and the Czech Republic. There is thus some limited experience of working with states outside the Community.

Other examples of river basin plans or accords covering international rivers in Europe are listed in Table 13.6.

Management plans for lakes

Lake management plans have generally been focused upon specific problems at individual lake sites. Lakes, particularly large lakes, are often used for water supply purposes, and those that are not tend to have important ecological or recreational significance to the local areas in which they are situated. Whilst lakes are clearly part of the river basin system as they receive water from a water catchment and are part of the natural drainage morphology of the basin as water flows into the lake or reservoir and passes through the water body into the river system, their management is often seen as a quite separate issue to management issues concerning the river itself. Consequently, although there are many examples of river basin management plans now in existence, these ignore the lakes, or separate them into a different category. Often, the land surroundng lakes and particularly reservoirs, is privately owned by water companies, canal companies or private landowners and this leads to problems in ensuring that the necessary management options are implemented. The Framework Directive changes this situation in that lakes and reservoirs must be considered as constituent water bodies in the river basin, and although their individual ownership may not be altered, plans must include reservoirs and lakes as ecological entities to be accounted for in any river basin plans.

The Kiskore reservoir on the River Tisza in Hungary is an example of a lake within a river system whose usage has changed in recent years and which now requires a management plan to meet its new function whilst also recognising its influence on the river basin. This lake was constructed in 1974 when irrigation water was required to support agriculture in the Tisza valley. Major socio-economic changes have taken place in the area since 1980 with the result that the reservoir, whilst maintaining importance for irrigation, now also functions as a source of hydropower water, and flood control. It is now also in demand as a major recreational facility, with angling, water sports and nature conservation as important functions. The reservoir is situated in the plains of the middle Tisza and complex rules govern its operation for flood control and in support of navigation on the Tisza. These have to take account of ice cover which occurs regularly for around 70 days a year. As the reservoir is allowed to drain in anticipation of flood waters, the resultant shallows can freeze completely causing the death of fish life. Multiple use of the reservoir is not in the best interests of individual uses such as angling and recreation. The river Tisza itself, which is the contributor of water, nutrient, pollutants and sediment upstream of the reservoir, is an international watercourse involving Hungary, Slovakia, Romania and the Ukraine. It is also a major tributary of the Danube, joining that river downstream of the reservoir. In order to draw up a management plan for the reservoir, river basin influences must be taken into consideration as the condition of the lake will be influenced by management decisions about the upstream rivers, and the way the reservoir is managed will affect the conditions of the river downstream (Petr and Olah 1997). In Turkey, the concept of catchment management plans for lakes has recently been proposed to protect six drinking water reservoirs serving Istanbul. In this example the reservoirs have been recognised as suffering from increasing problems due to urbanisation and insufficient supporting infrastructure. A catchment management plan, based on the establishment of protection zones around the reservoirs, has been proposed linked to new land-use plans for the areas. Such protection zones have a legal status in Turkey under a 1988 water pollution control regulation. The adoption of a land-use based plan, using the results of a prior survey to identify problems and possible solutions, enables controls to be placed on all types of activities that occur within the catchments, including agriculture, industrial discharges, human settlement development, recreation and so forth. Such an approach is applicable to uses other than water supply, and could be used for ecological protection (Tanik 2000).

In Scotland, a Loch Lomond Catchment Management Plan has been developed by East of Scotland Water, the Scottish Environmental Protection Agency, the Scottish National Heritage and West of Scotland Water. Alongside the Loch Lomond and the Trossachs Interim Committee, these four bodies form the core of a Steering Group for the project. The project officially began in November 1999 and aims to '...achieve environmental improvements... in a manner which takes into account socio-economic needs, while allowing for a greater understanding between concerned parties' (Frazer 2001).

The Loch Lomond area is renowned for its scenic quality, conservation value and status as a major tourist attraction. There are a large number of natural heritage designations in the area, including plans for the establishment for Scotland's first National Park. In addition, the impending EU Water Framework Directive and the possible designation of the River Endrick as a special area of conservation, have increased the pressure for a holistic view on management of the catchment.

Consultation in the Loch Lomond Catchment Management Plan aims to establish the main areas of concern within the area, involve the key players in the development of management proposals, and to produce an agreed action plan that can be effectively implemented and monitored over the next 5 to 6 years. There are four main stages:

Production of an issues report – through a process of consultation to establish to main issues of concern to those who live and work within the catchment. This report was completed in June 2000.

Production of a consultation report – this report has two main functions. The first is to provide an overview of the catchment, its environment and the activities undertaken within its boundary. The second section will contain the draft management proposals developed through the use of consultation groups. This report is due to be published in April 2001 and will be open to a 3 month consultation period. The results of this consultation process will be utilised to produce the third report of the project.

Production of an action plan – this final, agreed action plan will contain agreed management proposals and action points and will form the basis of the future work. It is due to be published in Autumn 2001.

Implementation and monitoring – it is essential that all the proposals contained within the action plan are achievable so that their effectiveness can be monitored to ensure that the project results in positive benefits.

It is hoped that adaptation of the plan to the needs of the EU Water Framework Directive and the establishment of Loch Lomond and the Trossachs as Scotland's first National Park, will mean that this Catchment Management Plan will continue to achieve its objectives for many years to come. The four stages discussed above are the route chosen to implement the Catchment Management Plan in Loch Lomond. However, each area has differing requirements and issues which can be addressed through the catchment management plan process. This is seen as one of its great benefits – it is an adaptable process that allows for environmental improvements at a local level. Catchment management planning meets the requirements of sustainable development, as it is an open, accessible and accountable process that works from the bottom up, and allows for debate and increased public participation.

International lakes

There are some lakes whose shores are shared by a number of countries for which international collaboration is needed where overall management of the lakes is required.

Lake Constance

The Agreement Concerning the Protection of Lake Constance is an example of an existing management plans for a cross-border lake. It was set up in 1960 as the Steckborn Agreement, and involves the Land of Baden-Wurttemburg, the Republic of Austria, Switzerland and Bavaria. The plan is focused particularly upon work to reduce the nutrient levels in the lake by extension to effluent treatment facilities and is implemented through an International Commission for Water Protection on Lake Constance. A series of further agreements regulate water abstraction and control shipping on the lake. In 1987 a set of guidelines was issued by the Commission to aid those organisations concerned with management of the lake.

The Great Lakes

Looking further afield, an early and significant example of international co-operation over lake management was the agreement enjoined between the United States and Canada over the management of the Great Lakes. Agreement was first reached in 1972 and amended in 1978, with further extensions to deal with the problems caused by phosphorus loading in 1983 and amended by a more substantial protocol signed in 1987. The agreement is a commitment by two sovereign countries to: control the discharges of toxic substances into the lakes; construct publicly owned wastewater treatment facilities; coordinate planning processes; and develop best practices for pollution control. The agreement is supported by legislation in both countries. In the context of the EU Water Framework Directive the agreement contains a number of similar concepts and is a useful example of how such a system could work between countries in Europe.

The main Protocol of 1987 sets out a number of procedures that are very similar in concept to those of the EU Water Framework Directive. The protocol sets out its purpose as being to restore and maintain the chemical, physical and biological integrity of the Great Lakes Basin ecosystem through programmes, practices and technology and to reduce or eliminate discharges of pollutants into the system. The policies include the virtual prohibition of the discharges of toxic substances; the development of coordinated planning processes; and best management practices. The protocol goes further than, and differs from the Directive, in that it aims also to provide financial assistance for the construction of appropriate publicly owned waste treatment works. This is not in accord with the polluter pays principle adopted in Europe. Many aspects of the practical implementation of the agreement are reflected in the way that the Directive operates. For example specific water quality objectives are adopted. These cover general water quality in the lakes which are based on chemical substances including toxic chemicals but also those substances which cause eutrophication.

General objectives for the lakes are adopted in terms of water quality. Whilst emphasis is given to the achievement of chemical criteria, such as freedom from certain toxic chemicals, and a system of identifying "critical pollutants" - substances for which specific objectives have to be developed – is described, there is also acknowledgement that ecosystem objectives should be developed. Within the lakes, "areas of concern" have to be identified where the objectives fail to be met, and where there is beneficial use of the area or where the ecosystem is adversely affected. For these areas, remedial action plans and lake-wide management plans must be prepared embodying a systematic and comprehensive ecosystem approach to restoring and protecting beneficial uses. Within such plans information on the environmental problems must be fully described. The definition of the causes of the problems, evaluation of current remedial measures, an evaluation of alternative additional measures and a schedule for their implementation must be prepared. The identification of the agencies responsible for the work must also be made, and a description of surveillance and monitoring processes must be given. Consultation with other interested parties such as state and provincial governments must be undertaken, but there is less emphasis on public participation than in the EU Water Framework Directive. Whilst the emphasis is generally towards identifiable industrial and other sources of pollution, Annex 13 of the 1987 Protocol also deals with the issue of non-point sources of pollution, and the need to deal with these through land use control under watershed management plans. Contingency plans for emergencies are also covered by the agreement (Great Lakes 1987).

Plans for coastal areas

Coastal and transitional waters represent a different problem in planning terms. Transitional waters are clearly at the lower extremities of rivers, and in catchment management will normally be included in any plans for the river basin if this technique is used. Coastal waters on the other hand tend to be treated separately. The waters that are included under the terms of the Directive are limited in their seawards extent, and indeed cover considerably less breadth than territorial limits. Many countries monitor and look after their coastal waters and some have defined specific limits for responsibility for the pollution control agencies. The difficulty perceived for the requirements in the Directive is that the one nautical mile limit does not correspond to their pollution control or fishery limits, and such sea is not always identifiable with a river basin. New regulations will be required in most cases to define the new responsibilities and to designate into which river basin each stretch of coastal water will be placed for administrative control. There is a number of existing legislative instruments that require plans for certain coastal waters. Two specific EU requirements exist that require water quality plans – the Bathing Water Directive and the Shellfish Water Directive both set standards for water quality, and as a result regulatory authorities have needed to draw up improvement plans. The Urban Waste Water Treatment Directive also has an impact upon plans for construction or maintenance of waste water treatment facilities. However, these instruments do not necessarily relate such plans to river basins as required by the new Directive. It is true that countries such as the UK are in the process of identifying riverine issues as factors which influence coastal water quality, so are moving towards the relationship of river basins to the coastal environment, but this is at an early stage. Planning issues related to the coastal environment are largely concerned with development issues such as settlements, tourism and such like. There are, in a few cases, specific environmental issues but these are connected with sites of special scientific or ecological interest.

Coastal waters are part of the wider maritime environment. International action on sea pollution has had a long history and there is a large number of conventions and agreements in existence that attempt to control environmental quality in the seas and oceans. The principal conventions governing the coastal waters around the EU include several which have world-wide authority and some which apply only to the near European waters. In the first category are the Paris Convention and the Oslo Convention. The Paris Convention, signed first in 1974, covered the prevention of pollution of the sea from land based sources. It is relevant to the current Directive in that most inputs of polluting substances from land enter the sea via rivers and transitional waters and any river basin plan which is prepared will inevitably make reference to the relevant polluting materials. The *Convention on the Prevention of Marine Pollution from Land Based Sources* – its full title – was first brought into effect on 6 May 1978. The area of its operation is the North-east Atlantic, excluding the Baltic and Mediterranean Seas. Many of the EU Member States are signatories and the EU itself is a contracting party. In terms of its overall plan, the Convention set up a commission (The Paris Commission – PARCOM) and the Commission requires signatory countries to undertake work to reduce the input to the sea of a number of the most polluting substances. This includes the prevention of further contamination of coastal waters or the open sea, assessing the effectiveness of pollution reduction efforts, monitoring of the marine environment and reporting upon those results. PARCOM has adopted measures to reduce the discharge of mercury, cadmium, PCBs and biocides to sea, and has also made recommendations on nutrient reductions. It introduced an annual study of pollutants in 1988, which in 1990 represented a study of riverine and direct input of pollutants to the area of control. PARCOM has adopted the new EU approach to pollution control by concentrating on industrial processes rather than dealing with the polluting substances themselves and has recommended the adoption of the best available technique (BAT) approach to their control. The Oslo *Convention on the Prevention of Marine Pollution by Dumping from Ships and Aircraft*, which covers direct inputs of pollutants into the same sea area, was combined with the Paris Convention in 1992 as a *Convention for the Protection of the Marine Environment in the North East Atlantic* and signatories to this include Belgium, Denmark, Finland, France, Germany, Iceland, Ireland, Luxembourg, The Netherlands, Norway, Portugal, Spain, Sweden, Switzerland and the UK. There is a new plan for the sea area that incorporates the EU water policy principles of precautionary principle, polluter pays principle, and BAT (or its equivalent of best environmental practice and clean technology). A plan for the area has been adopted which included the establishment of a programme to assess water quality in the sea area, a reduction in discharges of dangerous substances and in nutrient discharges, elimination of the dumping of waste, and the collection of data on the sources of hazardous substances entering the sea. Work on defining BAT and best available techniques was set up. The results of several initiatives fit well with the proposals for planning to deal with coastal and transitional waters under the Directive, and the area covered is coincidental with much of the water framework eco-region approach.

In addition to the Oslo and Paris (OSPAR) Conventions there are three specific conventions that cover the three "enclosed" sea areas around the EU. The Helsinki Convention of 1974 (revised in 1992) covers the area of the Baltic, the Mediterranean Action Plan is set up to manage the Mediterranean Sea, and the North Sea Action Plan is in force to protect the North Sea from pollution. The signatories to these conventions reflect the mainly riparian states. They have similar aims to the OSPAR plans: to reduce input loads of dangerous substances and nutrients. Again, in the context of the new Directive, the activities of the Conventions and action plans will provide a background of information and actions that may be capable of assimilation into any river basin management plan which is drawn up for the transitional and coastal water bodies identified for the purpose of the Directive. The planning process for the new river basin district competent bodies should take account of the obligations and actions already undertaken in under the Conventions, and there will need to be a discussion and reconciliation of the various plans.

ECE Water Convention

An important general approach to the management of international rivers and lakes is set out in the ECE Convention on the Protection and Use of Transboundary Watercourses and International Lakes signed in 1994. This agreement obliged signatory countries to elaborate agreements between those who had shared rivers and lakes. The Convention sets out common principles of water management, many of which are reflected in the new Framework Directive. The general provisions include measures, preferably taken at source, to prevent control and reduce pollution of waters that may cause a cross-border impact, to undertake ecologically sound water management, with conservation of resources and environmental protection, to ensure reasonable and equitable use, and to conserve and restore ecosystems. These principles are common to the new Directive. Three of the basic principles of the EU water policy: the precautionary principle, the polluter pays principle, and sustainable use of water resources, are also a general provision of the Convention.

REVISION OF PLANS

The Directive places an obligation on Member States to review their river basin plans over a set timescale. The first plan must be formulated and published within 9 years from the entry into force of the Directive and must be reviewed within 15 years with further reviews every 6 years. The European Environment Agency uses a particular form of review which provides a system for analysing interrelated factors that impact on the environment. The system is known as the DPSIR framework. In a general environmental situation it operates by obtaining information on the Driving forces which affect the environment, the Pressures which emanate from such drivers, the State of the environment, the Impact of such pressures, and measures the Responses of these interrelated factors to changes in policies or in the situation as a result of investment or new regulations. This system is based on an Organisation for Economic Cooperation and Development (OECD) model. When compared with the information requirements of the new Framework Directive it is clear that this system is ideally suited to the review process for action plans. Whilst not a mandatory requirement, in view of the fact that all information concerning the implementation of the Directive will be accessed and used by the EEA to prepare reports under Articles 18 and 19, it will be useful to bear this system in mind when organising the review of river basin management plans. Within the planning cycle the DPSIR cycle fits rather well (although out of phase). A schematic of the cycle is shown in Figure 9.1.

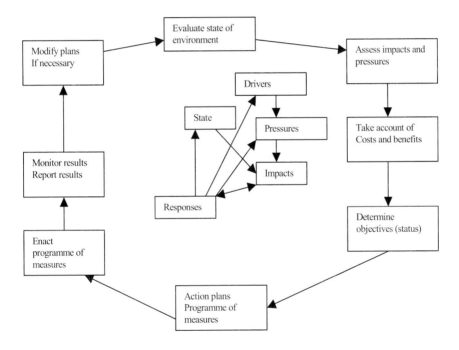

Figure 9.1 DPSIR cycle applied to the EU Water Framework Directive. (The outer circle is the planning cycle).

Timescales

There are deadlines set out in the Directive for the achievement of some of these stages. The Directive came into force in December 2000. By 2006 there must be a timetable for the production and consultation procedures of the river basin plans. Table 9.2 shows the activities to produce the plans which must be carried out by given dates and also indicates some of the related items which impact upon this process.

Table 9.2 Production schedule for the river basin plans.

Activity	Deadline	Related items
Issue timetable for production and consultation	2006	Completion of review of river basin characteristics by 2004
Publish interim overview of issues in each river basin	2007	
Publish draft plan for consultation	2008	Allow 6 months for comments on background documents used for development of draft plan
Publish river first river basin plan	2009	Issue the programme of measures for meeting objectives in the plan by 2009
Issue timetable for production and consultation for second river basin plan	2012	Initiate the programme of measures by 2012 Issue interim report on progress in implement-ation of programme of measures in 2012
Publish interim overview of new issues identified in second review of each river basin	2013	
Publish second draft plan for consultation	2014	
Review and update plan and issue second version	2015	Aim to achieve good water status in surface and groundwaters by 2015, except where extensions to the deadline have been made. Review effectiveness of programme of measures to achieve good status by 2015
Issue timetable for production and consultation for third river basin plan	2018	Issue interim report on progress of programme of measures under second plan in 2018
Publish interim overview of new issues identified in third review of each river basin	2019	
Publish third draft plan for consultation	2020	
Review and update plan and issue third version	2021	Aim to achieve good water status in artificial and heavily modified waters by 2021; Review effectiveness of programme of measures by 2021
		Issue interim report on progress of programme of measures under third plan in 2024
(Continue with further revisions to plans on a 6 yearly basis)		

REFERENCES

Correia F.N, Neves E. B, Santos M.A, Da Silva J.E. (1998) in *Institutions for Water Resources Management in Europe*, Correia F.N. (ed), Balkema, Rotterdam.

Danube (1994) *Strategic Action Plan for the River Danube 1995–2005*, Environmental Programme for the Danube River, 1994.

Frazer P., (2001), IWA Conference, Edinburgh.

Great Lakes (1987) *Great Lakes Quality Agreement of 1978, as amended by Protocol 1987*. Publ. International Joint Commission, US and Canada.

Petr, T., Olah, J., (1997) Multiple Function of Kiskore Reservoir and its Management Challenges, IWA Specialist Conference, Prague, 19–23 May 1997.

Regulation (1984) General Administrative Regulation Guidelines for Establishing Water Management Framework Plans, Bonn, 39 May 1984.

Tanik, A., (2000) Using the example of Istanbul to outline general aspects of protecting reservoirs, rivers and lakes used for drinking water abstraction, in *Water Sanitation and Health*, I. Chorus (ed), IWA, London.

Taw 2000 *Local Environment Agency Plan, Taw Action Plan from February 2000 to February 2005*, Environment Agency, Exminster.

10

Groundwater

GROUNDWATER PROTECTION

The EU Water Framework Directive is based firmly upon the integrated approach to water management. It acknowledges that the quantitative and qualitative status of groundwater can influence surface water ecology and therefore groundwater should be regarded as an essential, indivisible part of the hydrological cycle. However, there is also a view expressed in the Directive that groundwater is different in certain aspects to surface water. It cannot be seen; it may be polluted without realisation by the polluter or by the users; it has virtually no self-purification properties; and once polluted it is very difficult to return to an unpolluted state. This has impacts upon the surface water ecology where groundwater enters the surface water system (for example as a base flow for rivers), and it is very important in cases where the water is used as a source of drinking water or for other uses such as irrigation. Paragraph 28 of the preamble recognises that *the task of ensuring good status of groundwater requires early action and stable long-term planning of protective measures owing to the natural time lag in its formation and renewal.* The preamble indicates that these problems should be taken into account when establishing measures for the achievement of good status and when action is contemplated for preventing increases in pollutant concentrations. In view of this, despite the overall integrated approach, specific measures must be adopted to prevent and control groundwater pollution, and these are laid out in Article 17.

The Directive places an obligation on the Commission to identify Europe-wide measures for groundwater protection within a period of two years from entry into force of the Directive. The European Parliament and the Council of ministers must adopt the measures, which would then be introduced within Member States. The measures are aimed at achieving good groundwater status by 2015, 15 years after the Directive became law. The Commission has to propose general criteria which may be used to assess *good* groundwater chemical status, basing its views on the initial survey of river basin characteristics carried out under Article V, and the information on the chemical character of groundwater bodies found during the survey. It also has to propose criteria for identifying upward trends in the concentration of substances which would reduce the chemical status of the groundwater, and to define the starting point for measuring trends so as to ensure that measures to reduce pollution are leading to a reversal of any of these upward trends. Annex V, 2.4.4 requires that Member States use data from their monitoring programmes to identify long-term trends (in particular any upward trends in chemical parameters caused by human activities). Such trends need to be statistically justified with appropriate levels of confidence in the results. The Commissions proposals should deal with this aspect.

If the Commission fails to deliver the appropriate criteria within two years, then Member States must develop their own criteria, within 5 years from the entry into force of the Directive. The time scale for the Commission's exercise is extremely limited, as it is obliged to look at the results of each Member State's own surveys carried out under Article 5, and Member States have 4 years in which to complete these surveys, so that in theory the Commission has a year in which to make compete its work. It is likely therefore that, to develop Community level criteria, the Commission will have to use earlier data on groundwater quality if it is to produce proposals in time.

In the absence of such criteria, Article 17(5) permits Member States to adopt starting values of 75% of the levels of quality standards set out in Community legislation which already applies to groundwater. Such legislation includes the Directive concerning the protection of waters against pollution by nitrates from agricultural sources (91/676/EEC) and the directives concerned with drinking water in so far as this is untreated when delivered to the consumer. These include the Mineral Waters Directive 80/778/EEC on the approximation of the laws relating to the exploitation and marketing of natural mineral waters (as amended by Directive 96/70/EC), and the Drinking Water Directive 80/778/EEC as amended by 98/83/EC for public water supplies.

The Nitrates Directive sets a limit of 50 mg/l of nitrate in waters within nitrate vulnerable zones. The Mineral Waters Directive specifies limits for a wide variety of parameters for those waters which are recognised by the European Commission and the Member States as natural mineral waters (a complete list is published in the Official Journal from time to time). Where water is bottled without treatment for sale as drinking water, an increasing tendency in all Member States, the water must meet in all respects the requirements of the Drinking Water Directive. This instrument lists maximum legal limits for a wide range of chemical and bacteriological parameters – some 39 chemical and physical parameters, 11 toxic substances and 5 microbiological organisms are included. Where groundwater is abstracted and delivered direct to the consumer without the benefit of treatment, the water as delivered must meet these requirements. Effectively, these values therefore become the standards that must apply to the groundwater bodies. In the absence of other standards set by the Community, values for each of the named parameters set at 75% of the levels in the Drinking Water Directive become the default values of the water quality objectives for good status groundwater.

The measures adopted at Community level must be adopted by Member States and included in their programmes of measures under Article 11.

Groundwater Directive

There is already one measure which has been agreed at Community level. Article 11(3) contains a number of actions which will replace the current Groundwater Directive. Article 11(3)(j) effectively replaces the complex set of prohibitions and authorisations decreed under that Directive with a simple ban on the direct discharge of pollutants into groundwater. The Groundwater Directive (80/68/EEC) was introduced as a means of implementing Article 4 of the Dangerous Substances Directive (76/464/EEC). Its purpose is to prevent the pollution of groundwater by substances belonging to the families and groups of substances identified under the Dangerous Substances Directive as being particularly toxic, persistent in the environment, and which are known to bio-accumulate. The Groundwater Directive prohibits the direct and indirect discharges into groundwater of the so-called List I substances, except where groundwater is permanently unusable, and requires prior investigation of activities that might lead to discharges of List I and List II substances. It requires an authorisation procedure to be introduced for the discharge of List II substances. The lists contains the substances shown in Table 10.1.

Table 10.1 List I and List II substances for groundwater

List I substances	List II substances
Organohalogen compounds and substances that may form such compounds in the aquatic environment	Metalloids and metals and their compounds: zinc, copper, nickel, chrome, lead, selenium, arsenic, antimony, molybdenum, titanium, tin, barium, beryllium, boron, uranium, vanadium, cobalt, thallium, tellurium, silver.
Organophosphorus compounds	Biocides and derivatives
Organotin compounds	Substances that affect the taste of odour of or that cause the formation of such substances in groundwater and render it unfit for human consumption
Substances possessing carcinogenic, mutagenic or teratogenic properties in or via water	Toxic or persistent compounds of silica or those that cause the formation of such substances excluding those that are biologically harmless or are converted to harmless substances
Mercury and its compounds	
Cadmium and its compounds	Inorganic compounds of phosphorus or elementary phosphorus
Mineral oils and hydrocarbons	Fluorides
Cyanides	Ammonia and nitrites

Although the remainder of the Dangerous Substances Directive continued to be developed through a series of daughter Directives which covered individual substances and 17 elements and compounds were brought under control by the issue of specific emission limit values or quality objectives for surface waters, it was left up to Member States to determine the procedures for controlling the discharge of dangerous substances to groundwater. The implementation of this Directive has been largely achieved through the usual pollution control regimes covering the issue of authorisations for discharges into ground or through waste management licensing, or through the controls placed on discharges or activities carried out by industry under prior authorisation schemes such as the IPC regime in the UK or the Classified Installations Procedures in France. The wide-ranging nature of activities which enter groundwater (such as the disposal of pesticide residues from agricultural use) sometimes caused significant delays in introducing suitable regulations.

One of the problems inherent in the Directive was the difficulty of determining which of the substances within the families of substances in List I should be retained in that list (and therefore banned from discharge into any groundwater) or which substances should be transferred to List II (and be allowed discharge under licence), and similarly, whether there were substances which should be transferred from List II to List I. The wide-ranging possibilities for inadvertent discharge through waste disposal activities also added to the confusion. The regulations introduced in the UK applied the Directive to industrial processes that deliberately discharge such substances onto land or into groundwater, or where there is a risk of this occurring, to agricultural activities involving pesticides, sheep dips and herbicides, to owners of underground tanks which store these materials and to other users who dispose of materials to land without a waste management licence. Furthermore, it appeared that the amount of investigative work required prior to discharge may have been inordinate when compared to the risks involved (DETR 1998).

Article 11(3)(j) simplifies this procedure by prohibiting the direct discharge into groundwater of pollutants. The Directive defines pollutant in a very wide manner. A "pollutant" is *"any substance liable to cause pollution, in particular those listed in Annex VIII"*, and "pollution" has the usual EU definition – *"the direct or indirect introduction as a result of human activity, of substances or heat into the air, water or land which may be harmful to human health or the quality of aquatic ecosystems or terrestrial ecosystems directly depending on aquatic ecosystems which result in damage to material property, or which impair or interfere with amenities and other legitimate uses of the environment"*. What does this mean in terms of groundwater protection? In due course the Groundwater Directive will be repealed – on 22 December 2013 – and after that date discharges which are made *directly into groundwater* and which contain a very wide range of pollutants will be forbidden. The prohibited substances must include at least those which are listed in Annex VIII (Table 10.2) which effectively replace the List I and List II substances of the earlier Directive.

Table 10.2 Indicative list of main pollutants

Organohalogen compounds and substances which may form such compounds in the aquatic environment
Organophosphorus compounds
Organotin compounds
Substances and preparations or their breakdown products which, through the aquatic environment route, can affect steroid, thyroid, reproductive or endocrine related functions or have carcinogenic or mutagenic properties
Persistent hydrocarbons and persistent and bioaccumulable organic toxic substances
Cyanides
Metals and their compounds
Arsenic and its compounds
Biocides and plant protection products
Materials in suspension
Substances which contribute to eutrophication
Substances which have an unfavourable effect on oxygen balance.

It is interesting to note that, in a UK study carried out in 1995 (EA 1997), pollutants affecting groundwater abstractions were found to include the metals arsenic, copper, chromium, a number of solvents , the hydrocarbons diesel, fuel oil and petrol, a number of pesticides and cyanide, ammonia and some substances which affect the oxygen balance.

In practice, the Groundwater Directive provisions will remain in force for the next 13 years, but during that time the identified authorised direct discharges will have to be re-examined for their content of the above substances and for any other properties which could be regarded as polluting, and their authorisation will have to be withdrawn over the period.

The Directive permits a small number of exceptions to the provisions of Article 11(3)(j). These include: the re-injection of water extracted for geothermal purposes; the injection of water which contains substances in it which have arisen as a result of mining or oil extraction and exploration, or into geological formations from which hydrocarbons have been extracted, and into formations which are not suitable for other uses; the return of water extracted water into mines, quarries or construction sites; the injection of gas into certain underground strata for storage purposes; and the discharge of small quantities of substances for scientific purposes. There is also a provision for the authorisation of civil engineering and construction works to take place in situations in which this would have direct contact with groundwater (and thereby possibly cause pollution) but Member States have to develop general binding rules to cover these operations. These exceptions cannot be permitted if this would compromise the achievement of the environmental objectives for the water under consideration.

MEASURES TO DEAL WITH WATER QUANTITY

The Directive introduces a significant control regime for water quantity regulation. The "basic" measures which are a requirement under Article 11 include measures to promote an efficient and sustainable use of water, together with controls over the abstraction of groundwater which includes a requirement for prior authorisation. A register of water abstractors has to be set up for both groundwater and surface water. Abstractions which have no significant impact on groundwater status may be exempted. If it is intended to augment groundwater bodies or introduce artificial recharge, again a prior authorisation system of regulation must be introduced. Safeguards must, of course, be introduced to ensure that the use of surface or groundwaters for artificial recharge or augmentation does not compromise the status of the source water.

Although a matter largely concerned with possible contamination, the re-injection of water which has been abstracted for geothermal purposes may also be re-injected into aquifers, but subject to a scheme of authorisation. The re-injection of pumped mine-waters or waters which have originated as a result of mineral exploration or oil and gas extraction or as a result of construction or civil engineering works also fall within the same rules. Many Member States already operate permit systems for abstraction from groundwater, and the issue of artificial recharge is carefully controlled in view of possible contamination, so such a requirement should have little impact on existing regulatory systems.

MEASURES TO SAFEGUARD WATER QUALITY

Within the "basic" measures of Article 11, a system to regulate point source discharges which may cause pollution must be established, by prohibiting their entry into water or to authorise them with specific controls such as emission limits. Such a system may be applied to both surface and groundwater in situations where the provisions of Article 11(3)(j) do not apply i.e. where the discharges are not made directly into groundwater but pass into the unsaturated zone or onto the surface of the land.

Diffuse sources must also be controlled through a system of prior regulation such as prohibition or by the adoption of general binding rules. Such systems, although general in nature, may be applied to those activities which can affect groundwater through percolation of pollutants through the ground. Examples are given in Chapter 8. Most European countries have been moving towards the implementation of the Groundwater Directive as a means of protection quality, and many have systems in place for prior authorisation of direct discharges. The control of abstraction is probably less easy.

There are several key issues which will have to be resolved before the new provisions can be enacted. Possibly the most important change will be related to the question of responsible ownership of groundwater. Traditionally in many countries there is a link between land ownership and water. Landowners either own the water beneath their land in a similar manner to mineral rights, or they have rights to abstract the water and use it in whatever way they require. In recent years, many countries in Europe have moved towards the replacement of traditional rights in favour of statute law, generally as a result of a need to ensure that water resources are preserved and equitably distributed. For example, in Spain a Royal Decree of 1986 took the view that underground waters would be in the public domain, and in order to abstract water from beneath their land, a landowner would have to apply to the authorities for a permit. In parts of the country where water is scarce, this has an impact on the value of land, and in order to avoid opposition to the transition from private to public ownership, the Decree gave extensive protection to the existing rights of landowners with a long lead time for the measure to become fully operational.

Such rights included the right for landowners to carry out activities on the land provided groundwater quality was not affected.

In France, underground water has belonged to the landowner who has absolute rights to abstract and use the water, even to the extent that this may derogate the rights of an adjacent landowner. However, this view is changing and in the most recent 1992 water law, although rights of ownership are retained water abstractions are subject to a permit system if the amounts are more than minor. Groundwater abstractions of more than $8m^3$ per hour must be declared to the authorities, and those over $80m^3$ per hour require formal authorisation. This is important in the context of river basin planning as it will be necessary to allocate water resources between water uses in order to protect the groundwater. In the Netherlands and the UK the right to use groundwater rests with the landowner, but the degree of use is subject to a rule that no derogation must take place to an adjacent source. In the Netherlands, if such a situation occurs the aggrieved party can have the abstraction stopped under the Civil Code or, if this is not possible because the water is used for public supply, can claim compensation. In the UK such a situation was earlier dealt with through the Water Resources Act of 1963 under which all abstractions required a licence. This has been superseded by the Water Resources Act of 1991. Under this Act abstractors are required to obtain a licence if they wish to drill a borehole and abstract water.

Groundwater protection zone concepts

In addition to the application of general pollution control regulations, many countries in Europe and elsewhere in the world already consider it necessary to take special precautions to protect groundwater quality. The concept of "zones of protection" has been developed. The concept involves estimating the risk that an activity taking place on the surface of the ground surrounding the underground water body will cause a deterioration in the quality of the water, and where the risk is unacceptably high, demarcating an area of land overlying the groundwater within which hazardous activities are prohibited or controlled.

Commonly in Europe three such zones are identified. Where there is an existing or proposed abstraction source such as a well or borehole, an inner zone is identified within which, generally, all activities on the surface that are not related to operation of the well are prohibited. This is to ensure that the wellhead is protected from damage and that direct and immediate influence on the quality of abstracted water from, for example, the ingress of bacteria or spillages of chemicals, is prevented. A second zone is often defined which is larger than the inner zone, of such a size that pollutants which could reach the abstraction point may be given the opportunity to degrade or dilute to a non-hazardous level before the water is abstracted. In these zones restrictions on the types of activity permitted may be applied. A third, much larger, outer zone is sometimes defined. This may cover the whole of the catchment area of the groundwater and allows the authorities to take action to restrict or control activities which may have a long-term deleterious effect on the resource overall. For heavily exploited groundwaters the two inner zones may be combined into a single protection zone, giving immediate and short term protection to abstraction points, whilst the outer zone protects the groundwater resource as a whole.

Although primarily aimed at activities taking place on the overlying surface, in some places, where saline intrusion is a problem for example, or where the protection is related mainly to water quantity, the zones may define areas in which the drilling of new wells and the abstraction of groundwater is restricted. Outside Europe, particularly in arid countries, such issues are treated very seriously. In Jordan for example there is a master plan that identifies groundwater basins or "balance areas" for groundwater management purposes whilst in Tunisia the water law ("Code des Eaux") establishes the right to designate zones in which water abstraction is strictly controlled or prohibited. These are primarily aimed at protecting water quantity but do not exclude activities to protect the quality of water.

Identification of protection zones

The identification of groundwater protection zones is based upon a variety of principles. These range from simple pragmatic approaches, for example the designation of an area of fixed radius around a well, to the use of sophisticated models predicting the time of travel of pollutants, bacteria and chemicals to the point of abstraction and their likely impact on the water quality after this time to determine a precise boundary within which activities, which lead to these substances, may be prohibited.

A common factor in many schemes is the estimation of the vulnerability of the ground to passage of the materials through the ground, and various models of vulnerability have been developed using the characteristics of the geological strata and hydrogeological principles to assess water movement.

As a result of such work, a set of travel times for material to pass from the point at which it enters the ground to the abstraction point is derived. The sizes of the protection zones are generally based on a minimum number of days of travel or, in the case of the inner zone, often on a minimum distance. Despite the commonality of the problem in many European countries, zone sizes vary widely. Some idea of how the sizes of the zones vary is given by the examples in Table 10.3.

Table 10.3 Examples of protection zones in European countries

Country	Inner Zone		Middle Zone		Outer Zone	
	Time of travel	Radius of zone	Time of travel	Radius of zone	Time of travel	Radius of zone
Denmark		10 metres	60 days	300 metres	10–20 years	
The Netherlands	60 days	30 metres	(10 years)		25 years	
United Kingdom	50 days	50 metres	400 days			Whole catchment
Switzerland		10 metres	Individ-ually defined		2 x middle zone	
Ireland	100 days	300 metres				Whole catchment
Germany		10–20 metres	50 days			Whole catchment
Austria		<10 metres	60 days			Whole catchment

Land use planning

Property development of all kinds may affect groundwater quantity and quality by changing groundwater flow and from the production and use of polluting materials. Many countries utilise their land use planning legislation to regulate the construction of facilities that might have a detrimental effect on the environment and a number of countries are incorporating assessments of the likely impact on groundwater into their planning decisions. Groundwater can be protected by the careful application of land use policies. Land use planning is often a matter under the control of local authorities who control the siting of developments. Provided that the effects of the developments on groundwater is taken into account in planning decisions, this can be an effective tool for protecting water quality and quantity. There are several examples where formal action is taken to protect groundwater by this route. In South Australia for example land use activities which can degrade groundwater are prohibited or restricted by law under the Planning Act of 1982. Special "no-development" corridors of land are designated in Western Australia and in some parts of the US. In other countries such as the UK, formal consultation with the water or environmental authorities is necessary before permission can be granted for development under the Town and Country Planning Act of 1990.

Legal status of protection zones

Restrictions on the activities which may be controlled often have the potential to cause financial loss to the land occupiers, and the designation of groundwater protection zones usually requires legislation to ensure that any regulations are legally valid and can be applied and enforced within the zones. In most countries water law permits the establishment of statutory protection zones for groundwater.

For example in Germany water protection areas are set up under Article 19 of the water management law "in order to protect waters from harmful influences in the interest of the current or the future water supply, to feed the groundwater stock, to prevent rainwater from detrimentally flowing off, the soil from eroding, and fertilizers and pesticides from being leached into public waters" (Winter 1994). Activities such as intensive fertilizer use are prohibited and land owners have to tolerate regulatory activities such as water sampling. Definition of the boundaries of protection zone rely upon a code of practice. In Germany, compensation for financial loss is provided under certain conditions, although this is not always the case elsewhere. A similar situation exists in Austria, where a specific act allows the issue of an official notice that establishes protection areas in which prohibitions or limitations on land use may be introduced.

In the UK, Section 93 of the Water Resources Act 1991 allows the government to designate water protection zones where it is appropriate with a view to *"preventing or controlling the entry of poisonous, noxious or polluting matter into controlled waters"* (which includes underground waters), or *"restrict the carrying on in that area activities which are likely to result in pollution of any such waters"*. This section has been rarely used to date, and groundwater protection zones are designated using as a basis a nationally agreed policy operated by the Environment Agency but these are not legally binding except where such zones are identified as a result of specific EU legislation such as the nitrate vulnerable zones of the Nitrates Directive. Section 94 of the same Act specifically allows the government to establish such zones and to pay compensation where fertiliser use in restricted. Groundwater protection areas are used in an advisory capacity to influence the control of activities mainly through the land use planning route.

Outside Europe a variety of legal powers are used and groundwater protection zones take many different forms. In the US, for example there is legislation in some of the states that permits the establishment organizations specifically set up to manage protection zones. The organizations have been granted legal powers. In the state of Kansas, for example, groundwater management districts established under the Groundwater Management District Act can issue regulations setting out the minimum distances between wells, designate "no-development" corridors around alluvial formations, administer water rights issues and levy charges. In Arizona the 1980 Groundwater Management Act permits the establishment of Active Management Areas where groundwater is under threat in which a permitting regime is operated to control groundwater use.

In Australia planning legislation is used as the legal basis for groundwater protection zones. In the Perth area, for example, the by-laws of the Metropolitan Water Supply, Sewerage and Drainage Act 1909 are used to control those individual activities that have a potential to contaminate public drinking water supplies, whilst the land use planning legislation has been used to create a Rural groundwater catchment protection zone with legal status within which planning policy has to ensure that development is compatible with the long term use of groundwater for public supplies and ensure that land uses, which could have a detrimental effect on groundwater, are brought under planning control (Australia 2001).

Protection zones in the Directive

If the measures to protect groundwater include the establishment of protection zones, these will be recognised as protected areas under Article 6(1). The standards applicable to the water in the zones must therefore be met by 2015.

REFERENCES

Australia (2001) *Guangara Land Use and Water Management Strategy*, Western Australia Planning Commission.

DETR (1998) *Consultation paper – draft groundwater regulations,* Department of Environment Transport and Regions, London, 12 January 1998.

EA 1997 *Groundwater Pollution – evaluation of the extent and character of groundwater pollution from point sources in England and Wales*, Environment Agency, UK.

Winter, G., (1994) *German Environmental Law*, Martinus Nijhoff, Dordrecht.

11

Waters requiring special protection

PROTECTED AREAS

Some water bodies are designated for special protection under various legislative instruments. The Directive specifies in Article 6 that a register of protected areas must be established. Annex IV describes the protected areas. Article 6(1) specifies that the areas will have been designated *"as requiring special protection...for their surface water and groundwater or for the conservation of habitats and species directly depending on water"*. Thus, although the register is mainly concerned with waters that require protection, the land associated with such water may also be included in some instances. This requirement has a number of implications.

REGISTERS OF PROTECTED AREAS

First, a register of protected areas must be devised. No guidance is given on the format to be used in the registers, but it is clear that the registers must be able to distinguish areas according to river basins. The registers must contain maps and descriptions of the legislation under which the protected areas have been designated. If a national register is set up, for example, it must be possible to extract maps and information on a river basin basis, as the protected areas have to be included as a part of the river basin management plan. Once established, the registers must be kept under review and up to date. They must be set up within 4 years of the coming into force of the Directive, that is, by December 2004.

A number of countries already operate registers which could be adapted to the needs of this Article. For example, the Register of Vulnerable Zones under the Nitrates Directive, or the Register of Areas of Special Scientific Interest in the UK.

IDENTIFICATION OF PROTECTED AREAS

At first sight the content of the Register is limited to waters in a few special sites where EU legislation is already involved, for example, waters that fall into specific protection zones under EC Directives such as the Nitrates Directive. However, Article 6(2) extends the range of protected areas to a much wider definition. Perusal of Annex IV, which lists, for the purpose of this Article, the waters to which it applies, shows that a large number of waters have the potential to be included on the Register. Member States will have to take their own decisions based on this Annex.

Community water legislation

Specific community legislation includes waters that are identified under the Bathing Waters Directive (76/160/EEC). Annex IV also makes it clear that other recreational waters should be included. The possibility of including waters that are used for sailing, canoeing or other water related recreational pursuits must be examined. The new proposals for the review of the Bathing Water Directive, for example, may include a much wider definition of waters than those that are currently covered.

The two Directives that are concerned with nutrient levels in water also identify particular areas. The Nitrates Directive designates areas known as nitrate vulnerable zones. Under this Directive, where waters in these areas contain or may eventually contain nitrates levels above 50 mg/l, and where the nitrate level is caused by agricultural activities, action programmes have to be established to reduce nitrate levels. These may include using codes of good agricultural practice, or placing restrictions on the use of fertilizers. This Directive was adopted in 1991 and programmes should have been operational within 4 years. Since 1991 Member States have been engaged in implementing the investment programmes needed to meet the terms of the Urban Wastewater Treatment Directive. This Directive also contains provisions for designating areas that require special protection. They are the zones that contain waters subject to eutrophication, known as "sensitive areas". Again, such areas should have been identified by now, and action programmes set up to reduce nitrate and phosphate concentrations by the employment of nutrient removal processes at waste water treatment plants. Some countries in Europe have taken the decision to designate the whole of their territories and sensitive areas and vulnerable zones, and to set up management systems to deal with these particular problems on a national basis. There may be some problems associated with the input of material to registers of protected areas in such cases, as it will not be possible to associate particular areas with individual river basins.

Community nature legislation

A much wider issue is nature protection. The EU Water Framework Directive is primarily aimed at achieving good ecological status. The biological quality is thus very important. Nature protection fits with this concept, and there are already a number of EU and international instruments devoted to the protection of nature. In the context of the EU Water Framework Directive, whilst the emphasis is on water protection and the improvement of the aquatic environment many of the plants and animals protected rely upon satisfactory water status, and indeed the water environment is often essential to their survival. It is difficult to separate the protection of water from the protection of nature. Thus the sites designated under the Habitats Directive (92/43/EEC) and the Birds Directive (79/409/EEC) are essential sites to be included in the register of protected areas. These Directives are specifically mentioned in Annex IV.

The Birds Directive is linked with the 1971 Ramsar Convention on the protection of internationally important wetlands. Under this Convention large wetlands such as estuaries or inland water bodies, that are important to migrating birds, have been designated for special protection. Such wetlands will be clear candidates for inclusion on the register. Most countries in Europe have identified and registered wetlands under this Convention. Germany for example by 1992 had registered 29 such wetlands with a total areas of 671km^2, the largest area being the Wadden Sea.

The Habitats Directive requires "Special Areas of Conservation" to be identified, classified and designated and a management plan to be drawn up for the conservation of natural habitats and for protection of species. The main aim of the Directive is to promote the maintenance of biodiversity, taking into account social, economic, cultural, and regional requirements. The Directive divides Europe into five biogeographical regions – continental, alpine, Mediterranean, Macaronesian (Madeira, Canaries and Azores), and Atlantic, with 168 habitat types which are due for protection, listed in the Directive and including estuaries, lagoons and reefs. The number of species, for which protected areas must be designated, extends to 632. The Habitats Directive is the source of information on sites which will form the Natura 2000 ecological network throughout Europe in which "favourable conservation status" will be achieved for habitats and species selected as being of European interest. The Birds Directive uses the term "Special Protection Areas" and these must be identified for the protection of certain rare and migratory species. These areas will also contribute to the Natura 2000 programme.

Examples of other protected areas

Under the Common Agricultural Policy certain areas of land may be designated as requiring special protection involving the use of agricultural production methods compatible with the protection of the environment and the maintenance of the countryside. Council Regulation 2078/92 (the Agri-environment Regulation) approves a zonal programme, and detailed implementation rules, which relate to the way that agriculture is conducted in these areas, are set out in Commission Regulation 746/96. In the UK, this has been implemented by the establishment of statutory Environmentally Sensitive Areas (ESAs) using regulations under the Agriculture Act of 1986. Areas in the UK have been designated as warranting special protection for reasons such as their landscape, historical importance, or wildlife. In these areas farming practices, in particular, have both led to the present environment, and are likely to influence their future. As a result it is the way that farming is carried out that is now regulated so that the unique features of the areas are retained and protected. Some 22 such areas are designated. Many of these areas are not directly associated with the water environment, but, on the other hand, several are specifically identified because of their links with water. For example the Upper Thames Tributaries ESA comprises mainly the flood plains of the Thames from Oxford to Kelmscott, and includes the flood plains of the tributary rivers Ray, Cherwell, Glyme, Evenlode and Windrush within an overall watershed area of around 2500 km^2 (NAFF, 1994). The aim within these flood plains is to conserve and enhance the diverse wildlife of the wet grasslands, and management options include the possibility of raising water levels to encourage nesting and over-wintering birds (Upper Thames, 1994). The Broads ESA similarly covers the extensive and unique wetlands area of the Norfolk Broads (SI 1992/54 as amended) and the Lake District ESA extends over the Lake District in the north of England. Although the main issue for such protected areas is agricultural, there is no doubt that the water environment is extremely important, and within the concept of the EU Water Framework Directive such areas must be included in the register.

The UK also designates other natural sites for special protection. This may be for the purpose of protecting particular animals and plants or for a general ecological situation, but it may also be for the conservation of particular geographical or geological features. The legislation used is the Wildlife and Countryside Act of 1981 (amended in 1985) and in England the *Sites of Special Scientific Interest* (SSSI) network is the responsibility of the nature organisation English Nature. There are 242 national nature reserves in Great Britain. There are also two marine nature reserves – the Isle of Lundy off the coast of Devon and the Isle of Skomer, off Scotland. The way that such protection areas operate is to require the owner or occupier to seek permission to undertake any operations which might conceivable damage the protected entity, be that some geological feature or a particular plant or animal species. SSSIs are therefore suitable areas for inclusion within the register of protected areas under the EU Water Framework Directive. The Special Areas of Conservation of the Habitats Directive have been drawn from SSSIs.

Many other countries in Europe designate special areas for nature protection beyond those that are necessary to fulfil their obligations under the Habitats or Birds Directives. For example, Sweden has a system whereby the 21 County administrative boards prepare nature conservation plans, and at the local level municipal councils have also been given powers to establish nature reserves. The central Ministry of Environment has little part in this process but must approve the final lists. The identification of Special Areas of Conservation and Special Protection Areas has also been delegated to the local level. The resulting high number of sites to be designated under the Habitats and Birds Directives was the result of such local interest.

In Portugal National Ecological Reserves are established by statute (Decree Law 93/90). These include coastal areas and freshwaters. Such areas are specially protected from excessive and harmful human activities.

A similar level of designation exists in the UK under the 1949 National Parks and Access to Countryside Act under which local authorities may designate local nature reserves (English Nature 1991). It is clear from the legislation that legal requirements to preserve flora, fauna and the conditions in which they live fit well with the principles of the EU Water Framework Directive. Furthermore, as such areas are very localised, they may easily be identified within the river basin concept. There is a case for including these areas in the register, even though there will be an administrative burden in so doing. There are over 2000 of such reserves.

The National Nature Policy Plan of the Netherlands designates a national ecological network. This comprises key areas, nature development areas and ecological corridors. Although these are non-statutory areas, it is left to local councils to designate them in land use plans as legally binding areas. The key areas are those in which the existing ecological conditions are of national or international significance, and measures for the protection of the environment, particularly in respect of surface and groundwater, preventing changes in the water regime that could cause an adverse effect are required. Nature development areas are those that offer realistic prospects of development of ecological value, and ecological corridors contribute to migration. Such areas appear to fit with the principles of the EU Water Framework Directive and could be placed on the register.

In an accessing country such as Poland, nature protection is taken very seriously already. There are valuable natural ecosystems in Poland's mountains that are protected by the establishment of National Parks, Landscape Parks and Nature Reserves in which activities are controlled on a legal basis through the 1991 Act on Nature Conservation. Of relevance to the water issue is the protection within these areas of spring water zones and spa waters. In the water environment a number of wetlands is protected under the Ramsar Convention, and the "Green Lungs" wetland areas in Northern Poland are protected nature reserves. The application of EU legislation will lead to more protected areas being established in countries such as Poland.

In the Czech Republic over 1 million hectares is subject to the protected areas status and in the Slovak Republic a similarly high figure of more that 800,000 ha is the subject of protection orders. The Central and East European countries (CEEC) tend to establish nature reserves and afford them special protection in order to maintain biodiversity. To illustrate the need for this type of action it has been estimated that between 1750 and 1986, 36 plant species became extinct in Hungary, 20 of these before 1950, nine between 1954 and 1975 and in the remainder in the 10 years between 1976 and 1986. Before 1950 the major impact on the natural environment was due to the regulation of waterways from canal building and flood defence works causing 70% of the loss from extinction. Since 1950 there has still been a 38% loss due to water activities (WCU 1991).

Waters used for the abstraction of drinking water

A special case of waters requiring protection is that connected with drinking water. Water that is used for drinking purposes requires protection from those substances or organisms that are hazardous to the health of humans, and protection from those factors that would cause a diminution of quantity to such an extent that the water source becomes unavailable for use. Both issues may eventually lead to health related problems in the community supplied.

Surface waters for drinking are derived from flowing rivers or the static waters of lakes and reservoirs. Underground waters are derived from groundwaters contained in geological strata, known as aquifers. The two situations present differing protection requirements.

Protection of surface waters

The community has specified quality objectives for drinking water abstracted from surface water sources in a Directive (75/440/EEC) *concerning the quality of surface water intended for the abstraction of drinking water in the Member States.* This Directive requires that Member States identify such waters at their points of abstraction, classify them into three water quality classes, and ensure that certain chemical and bacteriological standards are met. For each class, A1 A2 and A3 of surface water, an appropriate water purification treatment process must be applied. The Directive applies without distinction to national waters and waters which cross international frontiers. Management plans to improve the quality of A3 waters are required so that this category of water is significantly improved by 10 years after the Directive came into force i.e. by 1985. There have been many problems in applying this Directive, particularly concerning its relevance in view of the more stringent values laid down by the later Drinking Water Directive 80/778/EEC and its own amendment 98/83/EEC, which obliges Member States to apply water treatment processes which are sufficiently rigorous to ensure that the water actually supplied to consumers meets the standards. The Framework Directive repeals the Abstraction Directive seven years after the new Directive comes into force, replacing it with the more flexible approach of Article 7(3) under which protection must be given to water bodies from which water is abstracted for water supply so that the level of treatment needed to meet the Drinking Water Directive standards is reduced.

The means of protection for drinking water supplies are not specified. These can take various forms, but for surface waters should include the use of pollution prevention measures and such procedures as reservoir catchment protection plans described in Chapter 9 and the measures to prevent over-abstraction described in Chapter 10.

Groundwater protection

Groundwater is an important source of drinking water supplies throughout the community. Special protection must be afforded to groundwater and protection zones are described earlier (see Chapter 10). Article 7 requires the identification of all bodies of water providing more than 10 m^3 per day or serving more than

50 persons, or water bodies which are intended for such a level of future use. The Directive does not specifically deal with groundwater resources that are used for irrigation, nor with small sources of groundwater, although there are many thousands of such small sources and both irrigation and individual supplies warrant the same protection as major public supplies. Where small supplies originate from a larger aquifer, the identification of the appropriate body of groundwater is dealt with as a matter of characterisation of the river basin (Chapter 6). If the overall groundwater body is sufficiently large to allow the supply of 10 m^3 per day it should be automatically granted a protected area status. The existence of a zone of protection or safeguard zone, as used by many countries, may also be a reason for placing it on the register of protected areas, but the legal status of such a zone might have to be considered first.

Economically significant aquatic species

Annex IV introduces the concept of identifying areas in which water requires special protection because it contains economically significant species. This is an interesting idea, as it will be a matter for Member States to decide whether or not particular fish, bird, or other animal and plant species are economically significant. There are some obvious cases, where existing EU legislation recognises the economic importance of some species. For example the Shellfish Water Directive (79/923/EEC) recognises areas in which shellfish thrive, and sets quality standards. Its later health related Directive concerning the placing of live bivalve molluscs on the market (91/492/EEC), lays down standards for shellfish flesh which reflect water quality in the shellfish growing area and requires competent authorities to demarcate the areas of saline or brackish water from which such molluscs may be collected for direct sale. These Directives identify areas that require special protection. However, a number of other aquacultural activities may be relevant to the register. For example in the Scandinavian countries of the EU and the north of the UK salmon farming is a commercial enterprise. It is possible to identify the particular water bodies in which such activities take place. Whilst it is often the case that the activities themselves cause water quality problems from the use of fish food and prophylactics, nevertheless, from the point of view of this Article in the Directive, such areas maintain commercial aquatic species. The cultivation of watercress takes place in certain rivers and streams in southern England. Again, this is a significant economic activity in the areas concerned, so it would seem appropriate to identify such areas as requiring protection. None are so designated yet. But what of commercial inland fishing – fish farming is one activity where the economics are readily available in the form of turnover for commercial firms, but sport fishing in lakes and rivers is also very important as a tourist attraction for local areas. It may be that such waters would be regarded as protected areas under the Freshwater Fish Directive (87/659/EEC) but this Directive is to be repealed within 7 years in favour of the ecological protection approach of the Framework Directive, and it does not specifically identify protection areas as such. The interpretation of the need for protection areas is not so simple as could be construed by a simple reading of Article 6.

Coastal environments

In the UK 1500 km of coastline in 45 sites have been designated "Heritage Coasts" since the 1970s, primarily to conserve the natural beauty of the coastline. The aims of designation have been widened to include protection of particular habitats for flora and fauna so consideration is needed for inclusion of such sites within "Protected Areas". Some of the sites already qualify due to their designation under such instruments as the Birds Directive. In Scotland there are 26 "Preferred Coastal Conservation Zones" covering 4700km that may also qualify.

The EU, and several of its individual Member States, are signatories to the 1982 Protocol concerning specially protected areas in the Mediterranean Sea, which was developed out of the Barcelona Convention of 1976. This Protocol established particular zones of the Mediterranean as being sites of ecological and biological value, with special interest for genetic diversity value and which are representative of types of ecosystems and ecological processes. In the context of the EU Water Framework Directive, if such sites fall within the coastal or transitional zones, they should be regarded as protected areas under this Directive. It is interesting to note, however, that the Mediterranean Sea is one eco-region for Directive purposes. In view of the fact that it is a virtually closed sea, collaboration may have to be sought under the terms of the Directive between Member States and States outside of the EU, in addition to the collaboration entailed through the Barcelona Convention (EC 1984).

RESPONSIBILITIES FOR PROTECTED AREAS

An issue that may arise in implementing the Directive is the question of responsibility for protected areas. In many countries such areas are designated and managed by a wide variety of organisations, government and non-governmental. The wide description of potential protected areas might involve ministries of environment, nature conservation councils, local authorities and municipalities, and often, private landowners. For example, in the UK the statutory non-governmental agencies are English Nature, Scottish Heritage and the Countryside Council for Wales. The National Park Authorities were set up in 1995 under the Environment Act 1995 to manage the many national parks in England and Wales. Nature Conservancy agencies – English Nature and Countryside Council for Wales in 1991, and Scottish National Heritage in 1992 – were set up to establish, maintain and operate national nature reserves, and to identify and notify SSSIs. The Environment Agency is responsible for non-statutory groundwater protection zones, and at least one statutory surface water protection zone. Landowners such as farmers are responsible for managing such designated protected areas as SSSIs, environmentally sensitive areas, and nitrate vulnerable zones; some water companies may control water source protection areas around the reservoirs that they own, or in upland catchment areas. At present, such zones are managed on the basis of their individual characteristics and needs, but in the future these zones will be included within the river basin management plans for the river basin district. At some point the accountability for activities within such protected areas will need to be determined in order to ensure that the river basin objectives are met in the timescales required by this Directive, rather than possible timescales imposed by the responsible bodies.

This problem is not unique to the UK. For example in Portugal the management of protected areas under the existing EU legislation lies with the Institute for Nature Conservation that is a central department of the Ministry of Environment and Natural Resources. Five Regional Directorates are responsible for all aspects of environmental management at regional level and coordinate all sectors including water and nature conservation. The areas they cover do not correspond with the main river basins of Portugal and their decision making processes are generally unrelated to river basins except for the requirement described earlier to prepare 15 specific river basin plans. The potential for problems in coordinating the needs of river basin planning in respect of the protected areas under nature legislation is likely to be high.

Further problems are foreseen where protected areas cross international borders. There is a need for close collaboration on river basin planning. The extent to which collaboration is needed for areas requiring special protection appears even greater. The identification of the responsible bodies, and their relationship to river basin management competent bodies, will be a significant item in implementation of the Directive.

OBLIGATIONS FOR PROTECTED AREAS

There are additional obligations in the Directive where protected areas are concerned. The monitoring programmes under Article 8 must take account of the additional requirements set out in the legislation that set up the areas if they are based upon Community law. So for example, the protected areas under the Abstraction for Drinking Water Directive must include within the routine sampling programme the specific requirements of the subsidiary Directive on methods of measurement and frequencies of sampling and analysis of surface waters (79/869/EEC). This may be a different requirement to domestic arrangements for water quality assessment of rivers and reservoirs. The Bathing Water Directive likewise has specific sampling programmes and methods that must be included.

Protected areas are subject to specific objectives. Article 4 requires that such objectives must be achieved at the latest by 22 December 2015 unless the Community legislation specifies a different date. There may be specific rights of extension of the time scale, or in some cases, reduction of the timescale set out in the legislation that established the protected areas. The legislation needs to be consulted.

Protected areas must appear in the river basin management plans identified on a map, with details of the legislation establishing them. A map of the monitoring points established for the protected areas, together with results of the monitoring programmes so far undertaken showing the compliance or otherwise with objectives, must also be supplied. The first river basin management plan is not due for publication until December 2009, but the monitoring programmes must begin by 2006, and furthermore, where the protected areas are already established under existing Community legislation, such programmes should already be under way.

REFERENCES

EC (1984) Council Decision 84/132/EEC, 1 March 1984, OJ L 68/36 10.3.84.
English Nature (1991) *Local Nature Reserves in England*, English Nature, London.
MAFF (1994) *Upper Thames Tributaries ESA*, Ministry of Agriculture Fisheries and Food, London.
Upper Thames (1994) *The Environmentally Sensitive Areas (Upper Thames Tributaries Designation Order*, Statutory Instrument 1994/ 712. HMSO, London.
WCU (1991) *The Environment in Eastern Europe 1990*, Environmental Research Series No 3, World Conservation Union.

12

Priority substances

DANGEROUS AND POLLUTING SUBSTANCES

One of the objectives of the Directive is to eliminate hazardous substances from the aquatic environment and to reduce the levels in marine waters to near background. So says the preamble to the Directive. References to the elimination or reduction of pollution occur at various points in the Directive as the achievement of good chemical quality is an essential part of ensuring that surface water ecology is satisfactory and that groundwater can be used as a source for drinking water. In the context of achievement of environmental objectives, Article 4(1) places an obligation on Member States to implement measures that will reduce pollution from "priority" substances and remove priority hazardous substances from surface waters and reduce pollution of groundwater by reversing any upwards trends in concentration of pollutants which are the result of human activity. Monitoring programmes established under Article 8 must be able to demonstrate the success of this action. The programme of measures, which must be put in place, must include measures to eliminate pollution from surface waters by certain priority substances identified in a procedure set out in Article 16, and by other substances that could prevent the environmental objectives from being achieved.

The actions in Article 16 are mainly directed towards the Commission in that the European Parliament and the Council of Ministers have to *adopt specific measures against pollution of water by individual pollutants or groups of pollutants presenting a significant risk to or via the aquatic environment*. The measures must be aimed at progressive reduction of pollutants and the elimination of *priority hazardous substances*. The measures are to be identified and presented by the Commission on a Europe-wide basis. Within the actions to be addressed by the Commission is a requirement to submit a proposal to identify those substances that are regarded as presenting a significant risk to or via the water environment. Article 16 revises the existing Dangerous Substances Directive (76/464/EEC) and its many daughter directives. Directive 76/464 is a framework Directive for the elimination of, or reduction in, the pollution of inland, coastal and territorial waters by particularly dangerous substances.

The Dangerous Substances Directive

In the Dangerous Substances Directive (76/464/EEC), in order to achieve the aim of eliminating pollution from the substances considered most dangerous to the aquatic environment, on the basis of their properties of toxicity, persistence in the environment and bioaccumulation, all proposed discharges that contain these substances have to be subjected to a process of prior authorisation. The authorisation will contain fixed emission limits, the values of which are decided by the Council of Ministers acting on recommendations from the Commission. Common limits will therefore apply throughout the Community. The limits refer to concentrations or maximum quantities that may be discharged. In some cases the limit values are determined according to industrial sectors. As an alternative to the use of such fixed emission limits, Member States have the option of applying quality objectives to the receiving water, the values of which are again determined at Community level, and calculating emission limits for discharges based on the achievement of the objectives. If the Member State decides to use this option it has to provide evidence to the Commission, through the use of an approved monitoring system, that quality objective values are being achieved and continuously maintained throughout the area where this option is taken. The Commission is obliged to review this situation every 5 years.

The list of dangerous substances to which these requirements apply is given in an Annex, and they are known as List I substances. List I comprises groups and families of chemical substances. In order to arrive at legally binding emission limit values or quality objectives, the Commission has to go through the procedure of adopting such values in a subsidiary directive – known as a daughter directive. So far the Commission has defined specific controls in this way for mercury discharges from chlor-alkali electrolysis plants (Directive 82/716), other mercury discharges (Directive 84/156), cadmium (Directive 83/513), hexachlorocyclohexane (Directive 84/491) and 14 organochlorine compounds in Directive 86/280/EEC (amended by Directives 88/347 and 90/415).

The Dangerous Substances Directive also identifies a second set of substances and groups of substances (known as List II substances) which, though less dangerous than List I, have a deleterious effect on the aquatic environment and therefore require control. For these substances Member States must establish appropriate quality objectives and draw up programmes to reduce pollution by applying emission limits to discharges into water, which are calculated in terms of the quality objectives. The Commission has not set any values for List II substances but require Member States to report on their actions.

Directive 96/61/EC on integrated pollution prevention and control (IPPC) is partly replacing and enhancing these provisions by introducing a prior authorisation requirement for a large range of industrial and agricultural premises, and setting emission limits on the basis of the use of best available techniques. In theory, the use of BAT as a regulatory tool should lead to the application, in many instances, of emission limit values which are more stringent than the requirements of the daughter directives to 76/464. In those cases where the operators of processes find it difficult to achieve, through BAT, levels that are compatible with the Dangerous Substances Directive limits, it is these limits that will become the default values. In such cases the processes would have to be closed down. The IPPC Directive came into force in October 1999 and will apply directly to new industrial sites. There is a short period of transition during which the original 76/464 Directive will remain in force even for those industries that will become subject to IPPC. The Dangerous Substances Directive still applies to industries or situations where dangerous substances may be involved but, because of size constraints or definitional problems, the industries concerned are unlikely to be subject to IPPC.

The development of fixed emission limits and quality objectives for all the families and groups of substances named in the Dangerous Substances Directive has been a very slow process. Individual chemical elements and compounds falling within the families and groups of substances in List I were identified in a Commission Communication to the Council in June 1982 (EU 1982). Of the 129 original chemical substances on List I only 17 have so far been dealt with. It is with this in mind that the Commission proposed a new approach to dealing with dangerous substances in the aquatic environment that has been incorporated in the new Framework Directive.

NEW APPROACH

Article 16 sets out the way that substances that are harmful to the aquatic environment will be dealt with in future. There are two features – a new way of identifying the substances that should be controlled, and a different approach to their control. Article 16 of the Framework Directive sets out a new procedure for identifying substances that warrant special controls, and will replace the provisions of Directive 76/464 within a period of 13 years from 22 December 2000, except that transitional arrangements will apply with immediate effect to two of the Articles in that Directive. These are the Articles in 76/464/EEC that establish limit values and quality objectives for the currently identified List I and List II substances. The second change is the adoption of the "combined approach" for control of dangerous substances, and indeed for other pollutants. This is described in Article 10 of the Directive and in Chapter 8 of this book.

THE IDENTIFICATION OF POLLUTANTS

Under Article 16 the European Parliament and the Council of Ministers have to adopt specific measures against pollution of water by individual pollutants or groups of pollutants presenting a significant risk to, or via, the aquatic environment, including such risks to waters used for the abstraction of drinking water. These must be aimed at reducing discharges, emissions and losses, (or for a group of substances known as "priority hazardous substances", ceasing or phasing them out altogether). The priority hazardous substances are part of a larger group known as priority substances. Thus Article 16 sets out a legal basis and new methodology for identifying substances that warrant control at Community level. There are effectively three lists of substances to which the Article applies.

In Article 2 (29) "pollutant" is defined as *any substance liable to cause pollution, in particular those listed in Annex VIII.* Annex VIII is considered to be an initial list of substances that should, in the Commission's view, receive attention. The list mirrors the list of substances, given in Annex III of the IPPC Directive, which must be taken into account when fixing emission limits under that Directive. A link is thus established between the implications of substances that affect the water environment and the main controlling legislation for industrial emissions. One of the key aims of IPPC is the principle of progressive reduction in pollution through the application of BAT, and the full implementation of the IPPC Directive becomes an essential part of the implementation of this Directive. The indicative list of main pollutants is given in Table 12.1.

Table 12.1 Indicative list of main pollutants

Organohalogen compounds and substances which may form such compounds in the aquatic environment
Organophosphorus compounds
Organotin compounds
Substances and preparations or their breakdown products which, through the aquatic environment route, can affect steroid, thyroid, reproductive or endocrine related functions or have carcinogenic or mutagenic properties.
Persistent hydrocarbons and persistent and bioaccumulable organic toxic substances
Cyanides
Metals and their compounds
Arsenic and its compounds
Biocides and plant protection products
Materials in suspension
Substances that contribute to eutrophication
Substances that have an unfavourable effect on oxygen balance.

Seventeen individual chemical compounds, which fall within the indicative list at Annex VIII, are already the subject of legislative control through transposition of the daughter directives of 76/464. In some Member States progress on addressing the identification and setting of quality objectives under List II of 76/464 has also been made and legal controls – quality objectives and associated emission limit values – have been set. Although the indicative list is intended to give initial guidance on the substances that should be considered early in any revised control procedure, these substances remain formally under regulatory control. For existing discharges there is a time delay whilst IPPC becomes operational. Full application of IPPC to existing installations will not be achieved until 2007 and until that date the transitional arrangements of the IPPC Directive must operate. Article 20 of the IPPC Directive requires that Articles 3, 5, 6(3) and 7(2) of Directive 76/464 remain in force so that there is a continuing legal basis for the application of emission limits and/or quality objectives for the 17 dangerous substances so far identified from List I. However, the EU Water Framework Directive repeals immediately Article 6 and replaces it with Annex IX. Annex IX transfers the legal status of the limit values and quality objectives for List I substances previously established under the Dangerous Substances Directive and its daughter directives to the new EU Water Framework Directive. A transitional provision at Article 21(3) also gives legal status to the quality objectives previously established by Member States under Article 7 of Directive 76/464 and permits Member States to use the new Directive if they need to establish further quality objectives. Annex IX lists the daughter directives of 76/464 and perpetuates the legal status of the limit values and quality objectives that have been established by them, re-enacting the repealed sections of that Directive.

The second list of dangerous substances – known as "priority substances" in the Framework Directive is likely to be much longer. This list will replace the 1982 list of 129 substances (subsequently extended to 132). The original list was drawn up using a number of criteria. The list included many substances that were particularly important to the marine environment and included a substantial number of chemicals that were the subject of international agreements, particularly between maritime nations. These included substances identified under the Paris and Oslo Conventions, and the North Sea Conferences.

Table 12.2 Composite list of dangerous substances from PARCOM, the North Sea Conference, and the Dangerous Substances Directive

Metals	Organochlorine compounds and pesticides		Nutrients and others	PCBs
Mercury	Hexachlorocyclohexane,	Endosulphan	Nitrate	Total
Cadmium	Gamma-HCH	Dichlorvos	Total nitrogen	PCBs
Copper	DDT	Fenitrothion	Total inorganic	PCB28
Zinc	Aldrin, Dieldrin, Endrin,	Fenthion	phosphate	PCB52
Lead	Isodrin	Malathion	Total	PCB101
Tributyltin	Trifluralin	Parathion	phosphorus	PCB118
Organotins	Trichlorobenzene	Parathion-methyl	Salinity	PCB138
Chromium	Trichloroethylene	Azinphos-ethyl	pH	PCB153
Nickel	Tetrachloroethylene	Azinphos-methyl	Suspended	PCB180
Arsenic	Hexachlorobenzene	Atrazine	particulate	
Boron	Hexachlorobutadiene	Simazine	matter	
Vanadium	Carbon tetrachloride	Mothproofing		
	Chloroform	agents		
	Pentachlorophenol	Dioxins		
	1,2 Dichloroethane			
	Trichloroethane			

In the new system, the Directive specifies that the Commission should propose a scheme for identifying priority chemicals using the data generated from other programmes of risk assessment, for example the risk assessment programmes associated with large production chemicals under Regulation

793/93, the Uniform Principles Directive for plant products (91/414/EEC), and Directive 98/8/EC, or by using a methodology devised for targeted risk assessment for aquatic ecotoxicity or human toxicity through the water route which has been in operation for a number of years. These, and other regulations which apply to the marketing of chemicals, produce data that can indicate environmental hazards and risks. By using the data related to aquatic ecotoxicity and human toxicity transmitted through the water environment, a link can be established with the other assessment programmes to cover a wider range of chemicals so avoiding duplication of effort in identifying problem substances. The EU Water Framework Directive requires a review of its first list of such chemicals after 4 years. Because the work carried out under the other legislation may not provide all the data required within this short timescale if full rigour is applied, the Directive permits the use of a shortened version which places more reliance on existing environmental evidence, in particular evidence that a substance is both toxic in the aquatic environment and that it has actually been found in the environment. This alternative methodology also allows the use of evidence of as to the large scale usage of the particular chemical to target substances which, perhaps, should be investigated at an early stage. At a later review stage, further evidence can be used to modify the list.

In view of the timescale, the Commission has invoked this latter methodology, and has used a priority-setting scheme. This is based on a simplified risk assessment procedure which itself is based on scientific principles that take account of the intrinsic hazards of the substances concerned, evidence of their presence in the environment and other data concerning their production and usage. Known as COMMPS (combined monitoring and modelling based priority setting) and involving many organisations beyond the Commission itself the procedure has resulted in the identification of a first priority list of 32 substances. The procedure is shown diagrammatically in Figure 12.1

The first list of priority substances was identified by reference to the chemicals already subject to Community-wide legislative control through their presence on a number of international and EU instruments including List I and II of 76/464, the North Sea Conference, the candidate substance list of the Oslo and Paris Conventions (OSPAR), EU Regulation 793/93 and Directive 91/414/EEC. In addition a large number of substances, for which environmental data were available, was also included. The first proposed list was issued as a decision proposal to the European Parliament and Council in February 2000 (EU 2000). The list contains 32 substances shown in Table 12.3.

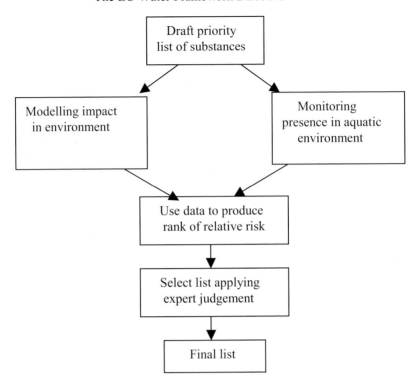

Figure 12.1 COMMPS procedure

Table 12.3 First list of priority substances

Alachlor	Anthracene	Atrazine	Benzene
Brominated	Cadmium and its	C10-13-chloroalkanes	Chlorpyrifos
diphenylether	compounds	Dichloromethane	Chlorfenvinphos
1,2-Dichloroethane	Di(2-	Hexachlorobenzene	Endosulphan
Hexachlorobutadiene	ethylhexyl)phthalate	Nickel and its compounds	Lead and its compounds
Mercury and its	Hexachlorocyclohexane	Diuron	Octyphenols
compounds	Naphthalene	Isoproturon	Polyaromatic
Pentachlorobenzene		Nonylphenols	hydrocarbons
Trichloromethane	Simazine	Pentachlorophenol	Tribytyltin compounds
	Trifluralin		Trichlorobenzenes

This list will serve as the first list of priority substances for 4 years after which it must be reviewed by the Commission and modified as necessary.

The next stage of this task for the Commission is to prepare proposals for Community-wide measures to reduce entries into the aquatic environment of priority substances from discharges, emissions and losses of the substances, paying particular attention to priority hazardous substances for which entries must be phased out over a timescale of 20 years from their identification. The detailed identification of priority hazardous substances has not yet been achieved. At the present time, the Commission has not come forward with its proposals for controls on these substances.

IMPACT ON MEMBER STATES

The immediate impact on Member States of this new system of control for substances which have serious impacts on the aquatic environment is to be found in Article 16(8). Under this Article, the Commission has to prepare and submit proposals for emission controls and environmental quality standards within 2 years of the placing of substances on the priority list. In order to do this it must take account of, and review, all of the directives listed in Annex IX for those substances which are included in the priority list and which appear in those directives, and possibly repeal those controls for substances which appear in those directives but no longer fall within the priority list. Annex IX contains the daughter directive of the Dangerous Substances Directive. The substances that currently appear both in Annex IX and the new priority list are: cadmium, mercury, hexachlorocyclohexane, trichlorobenzene, hexachlorobenzene, hexachlorobutadiene, chloroform, simazine, pentachlorophenol.

The important issue here is that if there is no agreement on Community-wide emission controls and environmental quality standards, Member States have to establish standards themselves. In the case of those substances for which daughter directives under 76/464/EEC have been issued, the quality standards are settled. But for the 23 substances that are now included in the priority list but which are outside of the dangerous substances controls, work must be put in hand to identify acceptable levels for these substances in the aquatic environment.

Toxicity assessment

The Directive lays down a procedure for this using toxicity assessment as its basis. The details are given in Annex V.1.2.6 of the Directive. Section 1.2.6 describes a methodology based on the determination of toxicity of the substance to three different types of aquatic organism – algae (or macrophytes), daphnia (or a representative organism for saline waters) and fish. The organisms must be relevant to the type of water body – that is, to the ecosystem classification determined during the initial survey of river basins. Once lethal concentrations and no-effect concentrations have been measured using standard test methods, in order to set an environmental quality standard the results are multiplied by a safety fact ranging from 10 to 1000 times.

Whilst many countries in Europe utilise toxicity testing for effluent assessment and for other purposes, the techniques are laborious and lengthy and this requirement will introduce a significant workload, particularly if there are many different ecosystem types recognised within their river basins, each requiring tests to be carried out on a range of organisms which are typical of the individual water bodies.

Once such environmental quality standards are proposed by the toxicologists, they must be subjected to peer review and public consultation to *allow a more precise safety factor to the calculated* (1.2.6(iv)). Such a consultation will add to the time taken to derive such environmental quality standards, and will probably lead to less rather than more certainty in the final values. The assessment of toxicity and the setting of environmental quality standards for all surface water bodies must be achieved by 2006, 6 years after the entry into force of the Directive. This is an extremely short timescale for such a huge task, taking into consideration the length of time needed to arrive at satisfactory environmental quality standards for the 17 substances covered in List I of the Dangerous Substances Directive.

Controls

The environmental quality standards, once set for each water body, will enable controls to be introduced on industrial and other activities which bear a relationship with the desired ecological status of the water, and these must also be in place by 2006 for those substances identified on the first priority list. The priority list has to be reviewed 4 years after the entry into force of the Directive, and every 4 years thereafter, and if there are any additions to the list environmental quality standards and the appropriate controls must be established within 5 years from the review date.

It is the Commission's job under Article 16 to prepare proposals for controls that lead to reduction of the impact of priority substances and particularly for the phasing out of the priority *hazardous* substances.

The controls which have to be introduced under Article 16 are likely to be similar to those adopted for other polluting substances. For example, the advent of the IPPC Directive gives a means of regulating emissions, discharges and losses from those industries that are the largest users of priority substances. The Directive points out that product controls under the Plant Products and Biocides Directives are appropriate ways of dealing with herbicides and pesticides and other items where controls of sale and use is relevant. The use of the combined approach is vital for countries to achieve adequate regulation and this is promoted elsewhere in the Directive. However, a new emphasis on the costs of so doing is an integral part of this process. The Commission, in proposing Community-wide controls, is obliged to seek the most cost effective and proportionate level of control, using a combination of means including product controls. As part of the preparatory work the Commission has to review the suite of Dangerous Substances Directives and decide whether the control of existing List I and List II substances is still warranted, particularly those substances that no longer feature in the new list of priority substances. Under Article 16 however the controls on such priority substances apply to all discharges rather than simply those that are covered by the IPPC Directive.

The Commission may also prepare strategies for dealing with other substances including their accidental discharge.

As a result of these strategies the overall level of pollution of water from dangerous and toxic materials should be reduced and water quality improved over a short timescale, leading in the longer term to a reduction in concentrations to background levels

REFERENCES

EU (1982) Communication from the Commission to the Council on dangerous substances which might be included in List I of Council Directive 676/464/EEC Official Journal C176 14.7.82 p 3.

EU (2000) *Proposal for a European Parliament and Council Decision establishing the list of priority substances in the field of water policy*, COM(2000) 47 final, Brussels 07.02.2000.

13

Monitoring

BASIC MONITORING REQUIREMENTS

Paragraph 32 of the preamble to the Directive identifies the need to monitor the water status on a systematic and comparable basis throughout the Community. This is necessary to ensure that the quality of the water is known, that there is a sustained interest in each Member State in the long-term maintenance and improvement in water status as shown by evidence of constantly improving environmental quality, and that each river basin is being actively managed in accordance with the Directive.

A corollary to this requirement is the need for comparability between results in time and place. Thus standardised methods which provide comparable results, whether the assessment is carried out in southern or northern states of the EU, must be used. Item 44 of the preamble calls for a committee procedure to ensure such comparability.

These needs are covered by Article 8 of the Directive – "monitoring of surface water status, groundwater status and protected areas" which requires the establishment of monitoring and sampling programmes within 6 years of the date of entry into force of the Directive. A great deal of discussion and negotiation has taken place regarding the appropriate ways of assessing water status. The result is expressed in Annex V which sets out in detail what should be measured in the context of surface running waters, lakes, estuarine and coastal waters and underground waters.

Three types of monitoring activity are envisaged: surveillance, operational and investigative.

Surveillance monitoring is defined in the Directive as monitoring to provide information to supplement and validate the initial assessment of the water status, assist in the design of future programmes, and to assess the long-term changes both from natural and human activities in the catchment.

Operational monitoring is to be carried out where waters may fall below their designated status, and are subject to management programmes. In this case the objectives of monitoring are more short term, and the programmes are looking for changes, both good and bad. The results may influence the direction of the programmes of measures.

Investigative monitoring is to be carried out in problem areas where the cause of problems in meeting environmental objectives is not known or where accidental pollution has occurred. Such monitoring will lead to measures to rectify the situation.

SURVEILLANCE MONITORING

Surveillance monitoring – surface waters

Although the Directive requires the identification of individual water bodies within each catchment, there is no absolute requirement to monitor each of these for surveillance purposes. Instead, in the case of surface waters, the Directive requires a programme to be set up within each catchment or sub-catchment of a river basin of sufficient intensity to establish an assessment of the overall water status. The main criterion for selection of sampling points is the quantity of water passing the sampling site. This should be significant in the context of the overall size of the river basin. There are additional requirements to sample at State boundaries and sites identified under the council decision 77/795/EEC on common procedures for exchange of information, but these should, by definition, be significant points. In due course this Decision will be repealed as it is overtaken by the new Directive.

The monitoring has to be carried out over a period of 1 year during each river basin plan period, so surveillance monitoring is normally a 1-in-6 year programme. If the water body is of good ecological quality and no changes have taken place this programme may be reduced to once in every three river basin plans. Annex V, at section 1.3.2, lists the parameters to be assessed. They include biological quality, hydromorphological quality, physico-chemical quality and measurements on pollutants (including priority list substances) as listed in Table 13.1.

Table 13.1 Parameters for surveillance monitoring

Biological parameters	Hydromorphological parameters	Physico-chemical parameters
Pytoplankton	Continuity	Thermal conditions
Other aquatic flora	Hydrology	Oxygenation
Macro-invertebrates	Morphology	Salinity
Fish		Nutrient status
		Acidification status
		Other pollutants
		Priority substances

The Directive recommends a sampling frequency for each of the programme elements, although fewer samples may be taken if there is justification for doing so. Biological and hydromorphological measurements must be made at least once during the sampling period. Physico-chemical measurements must be made at three-monthly intervals during each surveillance monitoring programme, and priority substances must be measured on a monthly basis. Care has to be taken to ensure that the times of the year at which samples are taken reflect the impacts of anthropogenic activity rather than simply showing the natural variation in the parameters, as the programmes are designed to show the changes caused by humans, whether good or bad.

OPERATIONAL MONITORING

Operational monitoring sites are aimed at demonstrating changes in individual water bodies. The sampling sites are therefore chosen on a different basis to surveillance sites. Monitoring points are specific to water bodies, and such monitoring is required only at those water bodies that are assessed as being at risk of failing to meet their environmental objectives or into which priority substances are discharged. The assessment is based upon the data collected during the initial survey of river basins, or from data that arises from the surveillance monitoring programme. The amount of monitoring for operational purposes may change during the course of a river basin plan as circumstances dictate. Annex V lays down more specific guidance for choosing the monitoring points. In particular, where legislation covers the discharge points of priority substances, these points must be included in the programme. In other cases, it depends upon the type of pressures to which the water bodies are exposed. If it is point sources, there must be sufficient sampling points in each specific water body to enable a proper assessment of the size and impact of the pollutant inputs. Diffuse sources are treated in a different way as, of course, their impact is more widespread and less definable. In this case monitoring points must be set up in a selection of water bodies, to establish the overall magnitude and impact of the pollutants, taking account of the risks of the diffuse sources causing a failure of environmental objectives. Hydromorphological problems are tackled in a similar way – using a selection of water bodies to assess overall impacts.

The quality elements which must be monitored have to reflect the pressures on the water bodies, rather than a fixed list of parameters, and include parameters that are indicative of the biological or hydromorphological quality which is likely to be most affected by the relevant pressures, and any priority substances which are discharged into the water. However, Annex V lists the following parameters as those similar to the parameters required for surveillance monitoring programmes.

The frequency of monitoring for operational purposes must be sufficient for the assessment of ecological status with acceptable confidence and precision. The level of confidence and precision has to be stated in the river basin plan, so the frequencies must be sufficient to be able to apply statistical techniques to the results. Annex V gives guidelines for the minimum frequency for each of these parameters, ranging from monthly analyses for priority substances, 3 months for most physico-chemical parameters, 6 months for phytoplankton and every 3 years for other biological organisms. But provided such minima are observed, it is up to Member States to choose an appropriate sampling frequency.

INVESTIGATIVE MONITORING

The precise form of investigative monitoring depends upon the situation. Its aim is to identify the causes of failures to achieve good environmental status, and to determine what measures are needed to improve or remedy the situation. Member States will have to decide how much, and in what form, such monitoring should be undertaken.

MONITORING IN PROTECTED AREAS

Some water bodies require monitoring at a level beyond the standard. Protected water identified under Article 6 may be subject to specific legislative or scientific requirements for data acquisition or for regulatory control. In particular surface waters, which are used for the abstraction of drinking water identified through Article 7 and supply more that $100m^3$/day, are subject to extra monitoring for priority substances and other pollutants which could affect their suitability as sources of water supply. Additional frequencies of sampling are set out in Annex V as in Table 13.2

Table 13.2

Size of source (Population served)	Frequency of measurement (per annum)
<10,000	4
10,000–30,000	8
>30,000	12

In practice it is unlikely that any water supplier would be sampling at less than this frequency in order to safeguard its position and meet the obligations of the Drinking Water Directive.

Operational monitoring is also needed to safeguard water quality and quantity in the other types of protected areas identified in Article 6.

METHODOLOGY

Annex V acknowledges the well-known problem of the difficulty of comparing like with like, when measurements are carried out using different techniques. The Directive requires Member States to use recognised international standard methods where these are available, or national or other international methods provided they give equivalent results both in terms of the quality of the results and their inter-comparability. Unfortunately, at the present time, few of the biological methods (for macrophytes, fish and algae) have not been developed to International Standards Organisation (ISO) or CEN standard, so reliance on national methods will be required. Inter-comparability exercises between Member States are underway until such times as the recognised standards are available. As described in Chapter 7 the Commission is organising an intercalibration exercise on classification using biological data, which will assist in smoothing obvious differences in interpretation, but this does not answer the need for specific scientific intercalibration of analytical and biological methods. In a project involving six of the countries through which the River Danube passes, an intercalibration exercise examining biological and chemical monitoring came to the conclusion that even though the countries involved were using nationally accredited methods, which were considered to give equivalent results, in fact there were significant differences between the country results (Danube 1998) that would be difficult to overcome, as each country considered that it was correctly applying methods of sampling, examination and analysis which were equivalent.

CURRENT POSITION OF SURFACE WATER MONITORING

How does this affect the current position of Member States and others with respect to their current activities? A study carried out on behalf of the European Environment Agency to determine the current level of monitoring examined surface water monitoring programmes in 16 countries of the EEA's area of jurisdiction (EEA 1996a). The purpose of the study was to obtain an overview of monitoring activities related to large bodies of water of interest at European and national or large regional level. To some degree the definition used in this study fits the surveillance monitoring specification proposed by the Directive. Annex V specifies that monitoring should take place at points where the rate of water flow is significant or where the volume of water is significant including large lakes and reservoirs.

River waters

The study showed that most countries have had such surveillance programmes in operation for a number of years, in some cases established through legal instruments, for example the Austrian "Ordinance on Water Quality Monitoring". Table 13.3 gives a short breakdown of monitoring programmes for major rivers in the EU.

Table 13.3 describes the basic situation with regard to physico-chemical monitoring programmes. There are, in addition, sampling programmes carried out to assess the biological status of rivers. In Austria, Denmark, France, the Netherlands, and Sweden, biological assessment is added to the general national river monitoring programmes, whilst in other countries specific monitoring programmes to assess biological river quality have been established. The activities are set out in Table 13.4

Table 13.3 Major river monitoring programmes: physico-chemical samples

Country	Date of establishment	Number of Sites	Approximate Annual sampling Frequency	Responsible Body
Austria	1991	244	6	Ministry of Agriculture and Forestry
Belgium (Flanders)	1989	957	8–12	Flemish Environment Agency
Belgium (Walloon)	1980	90	5	Environment Police Department
Denmark	1989	261	12–26	National Environment Research Institute
Finland	1962	68	4	National Board of Water and Environment
France	1987	1082	8	*Reseau National de Bassin*
Germany	1982	146	13	Joint Commission of Federal States
Greece	Early 1980s	~250	12	Laboratory of Soil Hydrology and Geology
Luxembourg	n/a	217	1–13	Administration of The Environment
The Netherlands	1955	26	6–52	Institute for Inland Water Management and Waste Water Treatment
Norway	1990	10	12	Institute for Water Research
Portugal	n/a	109	12	Regional Directorates of Environment and Natural Resources
Spain	1962	448	1–12	Central Commission for Water
Sweden	n/a	300	12	University of Agricultural Science
UK	1975	~10,000	6–52	Environment Agency /Scottish Environmental Protection Agency/Department of Environment (Northern Ireland)

Table 13.4 Biological assessment programmes for river waters

Country	Date of Commencement	Geographical coverage	Type of Analysis
Austria	1968	Nationwide	Saprobic index, Macro-invertebrates
Belgium	1989	Flanders region	Macro-invertebrates, Belgium Biotic Index
Belgium	1980	Walloon region	Macro-invertebrates Belgium Biotic Index, Phytoplankton Macrophytes
Denmark	1989	261 national sampling sites, with a further 10,000 local sites	Macro-invertebrates, Phytobenthos
France		National basin network 1082 sampling sites	Invertebrates, Fish
Germany	1976	All main flowing waters	Saprobic index, Macro-invertebrates, Microflora, Microflora
Ireland	1971	3000 sites in 1200 rivers	Macro-invertebrates, Macrophytes, Algae, Siltation
Luxembourg	1972	Main rivers	Belgium biotic index, Macro-invertebrates, Plankton, Macrophytes
The Netherlands	1992	Nationwide, 15 sampling Points	Macro-invertebrates, Fish, Phytoplankton, Zooplankton, Macrophytes
Spain	1980	847 national sampling points	Biological Monitoring Working Party (BMWP), Modified biotic index
Sweden	1993	35 streams	Macro-invertebrates, Periphyton
United Kingdom	Early 1970s	8266 national sampling points	Macroinvertebrates

Looking further afield to those countries seeking accession to the EU over the next few years, the picture is less clear. Most of the accessing countries appear to operate monitoring programmes for their larger rivers. For example physico-chemical programmes are in place in Poland, the Czech Republic, Slovakia, Romania, Hungary and most of the other States through which the River Danube flows, partly as a result of collaboration of those countries with the Danube Commission.

Lake waters

The Directive requires Member States to assess and monitor the ecological status of lakes within their area. This is an area of water quality assessment which, in general, receives less attention that that of the river systems.

  Some, but not all, of the EU Member States have established programmes to monitor their lakes and large bodies of still water. The position as at 1996 (EEA 1996a) is shown in Table 13.5.

Table 13.5 Lake sampling programmes

Country	Number of lakes	Sampling sites	Sampling frequency
Austria	No national programme except for large nationally important lakes – local monitoring elsewhere		
Belgium	No information		
Denmark	37	40	19 pa
Finland	>200	>200	3
France	No national programme – local monitoring		
Germany	No national programme – local monitoring		
Greece	Some monitoring included in river programme		
Ireland	530	Variable	Variable
Luxembourg	3	10	8
The Netherlands	Some monitoring included in river programme		
Norway	1356	>700	0.1– 12
Portugal	No national programme – some monitoring included in river programme		
Spain	No national programme – the inland monitoring programme includes some reservoirs		
Sweden	1100	N/a	0.2– 7
United Kingdom	No national programme – local monitoring of lakes, especially for blue-green algae		

International freshwaters

Annex V of the Directive also specifies that surveillance monitoring must be carried out on significant bodies of water that are passing from one country into the jurisdiction of another, even to those countries outside of the EU boundaries, and at sites at which it is necessary to estimate the pollutant load which is transferred across boundaries into the sea. This requirement is largely met by a number of international agreements which have been signed over the past four decades, that oblige Member States and others to undertake monitoring. In the case of surface freshwaters these comprise the following (EEA 1996c) (Table 13.6):

Table 13.6 International conventions relating to cross-border rivers and lakes

Name of convention	Countries affected
Foyle Fisheries Convention between Northern Ireland and Republic of Ireland 1952	Northern Ireland, Republic of Ireland
Convention on Co-operation for the Protection and Sustainable Use of the River Danube 1994 (Sofia)	Austria, Denmark, Bulgaria, Czech Republic., Slovakia, Slovenia, Hungary, Romania, Ukraine, Yugoslavia
Convention concerning water economy questions relating to the Drava, 1954	Austria, Yugoslavia
Agreement concerning water economy in respect of the frontier section of the Mure, 1954	Austria, Slovenia
Convention between Italy and Switzerland concerning Lake Lugano, 1955	Italy, Switzerland
Treaty between Hungary and Austria on water economy, 1956	Austria, Hungary
Convention between Baden-Wurttemberg, Bavaria, Austria and Switzerland on the Protection of Lake Constance against Pollution, 1960	Austria, Germany, Switzerland
Protocol for the Protection of the Moselle, 1961	Germany, France, Luxembourg
Protocol establishing an International Commission for the River Sarre, 1961	Germany, France
Convention concerning the Protection of Lake Geneva Against Pollution, 1962	France, Switzerland
Agreement concerning an International Commission for the Protection of the Rhine Against Pollution, Bern 1963	Germany, France, Luxembourg Netherlands, Switzerland
Agreement between Finland and the Russian Federation concerning Frontier Water Courses, 1964	Finland, Russia
Agreement between Germany, Austria, Switzerland on the Withdrawal of Water from Lake Constance, 1966	Austria, Germany, Switzerland
Treaty between Austria and Czech Republic on Frontier Waters, 1967	Austria, Czech Republic

Table 13.6 International conventions relating to cross-border rivers and lakes (*cont'd*).

Name of convention	Countries affected
Treaty on Frontier Waters between Austria and the Slovak Republic, 1970	Austria, Slovakia
Agreement between Finland and Sweden concerning Frontier Rivers, 1971	Finland, Sweden
Agreement between France and Switzerland concerning navigation on Lake Geneva, 1976	France, Switzerland
Rhine Convention against Chemical Pollution, Bonn 1976	Germany, France, Luxembourg Netherlands, Switzerland, EU
Rhine Convention against Chloride Pollution, Bonn 1976	Germany, France, Luxembourg Netherlands, Switzerland, EU
Agreement concerning cooperation on water management of Frontier Waters between Italy and Slovenia 1978	Italy, Slovenia
Agreement on a Finnish Norwegian Commission on Boundary Watercourses 1980	Finland, Norway
Agreement between the French Republic and the Swiss Confederation concerning the reduction of the phosphorus concentration of the waters of Lake Geneva, 1980	France, Switzerland
Agreement between Finland and the Russian Federation concerning the production of electric power in the part of the Vuoska River bounded by the Imatra and Swetogorsk Hydroelectric stations, 1972 and 1983	Finland, Russia
Bucharest Declaration (Danube), 1985	Austria, Germany, Bulgaria, Czech Republic, Slovakia, Slovenia, Hungary, Romania, Ukraine, Yugoslavia
Convention between France and Luxembourg concerning certain Industrial Installations on the Moselle, 1986	France, Luxembourg
Regensburg Agreement, 1987 (Danube Basin)	Austria, Germany, EU
International Commission on Hydrology – Rhine, 1989	
Agreement between the Republic of Finland and the Soviet Union concerning the release of water of lake Saimaa and River Vuoski, 1989	Finland, Russia
Elbe Convention, 1990 (Magdeburg)	Germany, Czech Republic, Slovakia, EU
Protocol for Technical Co-operation between Greece and Bulgaria, 1991	Greece, Bulgaria
Transboundary Watercourses and International Lake Convention, 1992	Austria, Belgium, Denmark, Germany, France, Finland, Greece, Italy, Ireland, Luxembourg, Norway, Netherlands, Portugal, Spain, Sweden, UK, EU
River Meuse Convention, 1994	Belgium, France
River Scheldt Convention, 1994	Belgium, France

Such agreements between states do not always deal with the issue of riverine discharges leaving a Member State's coastal boundaries, for which specific provision must be made. However, a number of international agreements require such activities, and have led to comprehensive monitoring programmes to meet this need, or in some cases the extraction of data from other monitoring programmes as inputs into the assessments of loads entering the seas concerned.

Transitional and coastal waters

The Commission is concerned about the impact that polluting inputs from riverine sources may have on coastal waters. In the preamble to the Directive, reference is made to the international agreements containing obligations to protect marine waters from pollution. It specifically makes reference to obligations approved by the Council of Ministers and which are therefore obligatory to Member States. The Directive and its monitoring requirements are seen as assisting the EU to meet these obligations.

Probably the most important of the agreements covering this area of monitoring is the Convention for the Prevention of Marine Pollution from Land Based Sources of 1974 – known as the Paris Convention. This covers the area of sea of the North East Atlantic including the North Sea, but not the Baltic or Mediterranean Seas, which have their own separate conventions. With the exception of Finland, all the countries party to this Convention have seaboards to this marine area. The signatories to the Paris Convention are Belgium, Denmark, Finland, France, Germany, Iceland, Ireland, the Netherlands, Norway, Portugal, Spain, Sweden and the United Kingdom. The Convention introduced two lists of substances that were to be brought under control, and a monitoring obligation in order to inform the administrators of the Convention (the Paris Commission). In 1992 a new Convention for the Protection of the Marine Environment in the North East Atlantic was agreed, the contents of which include a monitoring commitment to obtain information about land based discharges to the sea, including hazardous materials and nutrients. This Convention supersedes the Paris Convention.

Although located within the same marine area as that covered by the Paris Convention, additional monitoring of input loads to the North Sea is also undertaken by eight members of the Paris Convention (Belgium, Denmark, France, Germany, Netherlands, Norway, Sweden, the United Kingdom) under obligations agreed at a series of conferences between 1984 and 1995. These are the riparian states of the North Sea.

The other main marine areas bordering on the EU, the Baltic Sea and the Mediterranean Sea, are covered by two further Conventions – the Helsinki

Convention of 1992 (Baltic Sea) and the Barcelona Convention of 1976 (Mediterranean Sea). The latter has associated with it Protocol for the Protection of the Mediterranean Sea Against Pollution From Land Based Sources, signed in 1980. This contains rather similar provisions to the Paris Convention.

A surveillance monitoring programme is also required under Council Decision 77/795/EEC of 12 December 1977 (Exchange of Information about Surface Fresh Water) under which each Member State has to design a sampling programme to sample a number of rivers at designated sites, usually on a monthly basis, analyse the water samples for a specified number of parameters, and forward the results to the Commission. The initial set of sampling points was specified in Annex 1 of the Decision. An indication of the disposition of sampling points is given in Table 13.7.

Table 13.7 Sampling under Council Decision 77/795/EEC

Country	Number of Sampling Points
Germany	11
Belgium	9
Denmark	4
Greece	6
France	16
Ireland	4
Italy	16
Luxembourg	1
Netherlands	13
Portugal	12
Spain	15
United Kingdom	17

From the above resume of large scale monitoring effort it is clear that all countries that are current Member States of the EU undertake some form of surveillance monitoring of surface freshwaters. Article 8 of the Directive emphasises the river basin approach to water management by requiring the monitoring to be capable of providing a "coherent and comprehensive overview of water status within each river basin district". Some of the monitoring programmes described above may provide data that can be used to identify conditions within the river basin, provided the sampling points are representative of the conditions within the river basins, and sufficient sampling points within each individual basin are established to assess the water status conditions in enough detail for a picture to emerge. In order to use all of the sampling points it will be necessary to link data collected under several different sampling programmes. This will also require control of data quality.

The European Environment Agency (EEA 1996d) has proposed a network of European sites which will also move towards the requirements of the Directive. Such a network would be based upon sites already identified in national monitoring programmes supplemented by additional sites where necessary. The network would be a representative sub-sample of the inland water bodies of the EEA area, and the sites would be selected so that they are representative of the size, numbers and types of water bodies in the area, the variation of human pressures, and they should include a number of reference sites. The network has not yet been established.

MONITORING – GROUNDWATERS

Because the Directive is concerned with the overall water situation, groundwaters merit a similar degree of attention as surface waters. Sampling programmes for ascertaining the status of groundwaters are therefore an integral part of the Directive. Article 8 specifies that both the quantity and quality of groundwater should be taken into account when assessing the status of groundwaters. Groundwater is used generally as a source of drinking water and monitoring programmes need to take this into account. Annex V specifies that a groundwater level monitoring network should be established capable of assessing the available groundwater resource.

Measurement of quantity

The EEA (EEA 1996b) has identified a number of water level monitoring programmes which have been in operation in some cases for many years. There is considerable variation in the number of sampling points used, which range from a few hundred to many thousands, depending upon factors such as the degree of exploitation of a groundwater resource and the ready availability of sampling wells or boreholes. Table 13.8 illustrates the situation.

Overall, the EEA found that groundwater quantity monitoring has been undertaken by most countries of the EU for many years, with average records being available for the last 20–35 years of operation. Such monitoring has been, and is, undertaken to assist in groundwater management, particularly in respect of the safeguarding of water supplies, or for more basic geological investigations. In some cases it is carried out to ensure that over-abstraction does not allow saline intrusion to occur. It is rare, however, for there to be a legal obligation to monitor water quantity. Only seven countries monitor as a legal requirement under their water laws.

Table 13.8 Groundwater quantity monitoring networks

Country	Quantity Monitoring Points	Area Covered Sq km	Date Commenced	Responsible Organisation
Austria	3100	83,850	1930	Hydrological Service
Denmark	200	43,216		Geological Service of Denmark and Greenland
Finland	550	-	1968	Finnish Environmental Institute
Germany	71000 (Bavaria)	71,000	1915	*Lander* Authorities
	85901 (Nordrhein Westfalen)	34,005	1909	
	1225 (Thuringen)	16,340	1915	
	2400 (Sachsen–Anhalt)	20,000	-	
Greece	Individual special projects			Institute of Geological and Mining Research
Iceland	Drinking water wells	100,000		National Environment Authority
Ireland	Local wells	18,870		Geological Survey of Ireland
The Netherlands	4000	36,000	1870	TNO Institute of Applied Geoscience/Local Authorities
Norway	67 National points 20000 wells		1901	Geological Survey of Norway
Portugal	620	31,500	1970	Water Institute
Spain	6576	173,030	1967	Lan Nationoal de Gestion y Conservacio de Acuiferos (National Plan for Groundwater Conservation) (PNGC)
Sweden	357		1955	Swedish Geological Survey
United Kingdom	5418 (England and Wales)		1845	Environment Agency

The control of monitoring networks and programmes tends to be nationally based, although the German *Lander* and the French regions have autonomous control of their programmes. The measuring point density is extremely variable, from 7.3 sites per km^2 in Finland down to 0.004 sites per km^2 in Norway. Networks involve a range of different types of observation sites, from wells to boreholes, but nearly all countries use observation wells fitted with recording equipment for the purpose of groundwater level measurements. A national groundwater monitoring network is being established in some countries. For example, the UK aims eventually to have in place a network of 3000 monitoring boreholes at a mean density of 1 per 25km^2. The EEA is expecting the establishment of a European wide monitoring network in due course. Annex V also specifies that, where groundwater crosses international boundaries, sufficient measurements must be taken to be able to estimate the direction and rate of flow across the boundary. This will require special attention when Member States draw up their river basin plans for river catchments which extend across their boundaries.

Quality monitoring

The Directive requires surveillance and operational monitoring of groundwater quality. The surveillance programme is aimed at supporting the impact assessment carried out under Article 5 and to identify long-term trends in water quality. The most important criteria for the selection of monitoring sites is to ensure that waters at risk of deterioration due to human activity are adequately covered, and that water crossing international boundaries is sampled. The Directive gives a set of key parameters, but others must be added to this list as necessary to cover known impacts and to address the needs of potential uses in other countries where water crosses the State boundary. The key parameters are oxygen content, pH, conductivity, nitrate and ammonium. Operational monitoring is carried out for similar reasons as for surface water – to detect long-term upward trends in pollutant concentrations and to establish the chemical status of water, which are at risk of failing to meet the objectives. Monitoring sites must be chosen to reflect this, concentrating on the "at risk" groundwaters, but care must also be taken to ensure that the chosen points are representative of water in the groundwater body. The frequency of monitoring is at a minimum of once per year, but sufficient to demonstrate the impact of pressures.

Current position

The EEA's study (EEA 1996b) of groundwater monitoring showed that, although monitoring of groundwater quality has been undertaken in most EU countries for the past 30 years, there are many national differences in the extent an character of groundwater monitoring programmes related to monitoring obligations in national laws. Only France has had an very long established, formal, quality monitoring network. The number and types of sampling sites and the frequency of sampling, the parameters measured, the limits of detection used, and the quality assurance methods and statistical treatment of results, varied widely although the networks generally cover the entire territory of the State (in some cases organised regionally as in the German *Lander*) and are distributed throughout the groundwater areas although many are concentrated around drinking water abstraction sites. In some countries, as in the UK, the amount of information is enhanced by using data obtained by water suppliers as part of their quality assurance programmes for public health purposes and from samples taken under the Drinking Water Directive. There are many organisations concerned with quality assessment. In some countries, special interest in, for example, agricultural pollutants such as nitrates (the Netherlands) or acidification from air pollution (Denmark) has led to a concentration of monitoring activity aimed at this particular facet. In Portugal, Spain and the UK saline intrusion is a problem so the coastal areas have had special attention.

The parameters examined generally exceed those key parameters laid down in the Directive, and are generally adapted to national requirements. Again, much of this reflects concern over drinking water quality, as groundwater is often supplied to consumers in an untreated state, so such parameters as pesticides, chlorinated solvents, and heavy metals are often included. The methods of analysis and their precision and accuracy are based upon national standards rather than EU ones, although many of these are equivalent or the same as ISO/CEN standards.

Impact of the Directive

In order to implement the requirements of the Directive for groundwater monitoring, the most important activity will be the establishment of a network of sampling points to enable representative samples to be obtained in such a way that the long-term quality trends in groundwater bodies is identifiable through the surveillance monitoring programme, and that sufficient monitoring points are set up in water bodies identified as being at risk to be able to detect changes brought about by the programmes of measures introduced in the river basin management plans. Some countries will be able to identify sufficient suitable points from their existing networks. Others, particularly where the former activities have been largely aimed at waters abstracted for drinking or irrigation uses from existing wells and boreholes, may have to drill observation wells in order to extend the coverage to a sufficient degree. Chilton (Chilton and Milne, 1994) describes the development of a monitoring strategy based on representative observation wells at a density of one per 25km^2 for England and Wales to be introduced over a period of around 5 to 10 years. This may be the way forward for other countries.

The issues related to the standardisation of sampling and analysis techniques are common to surface and groundwater and will require a programme of standardisation over a period of years.

REFERENCES

Chilton P.J., Milne C.L., (1994) *Groundwater Quality Assessment: A National Strategy for the NRA*, Report WD/94/40C, British Geological Survey, Wallingford.
EEA (1996a) *Surface Water Quality Monitoring*, Topic Report 2, EEA, Copenhagen.
EEA (1996b) *Groundwater Monitoring in Europe*, Topic Report 14, EEA, Copenhagen.
EEA (1996c) *Requirements for Water Monitoring*, Topic Report 1, EEA, Copenhagen.
EEA (1996d) *European Freshwater Monitoring Network Design*, Topic Report 10, EEA, Copenhagen.

14

Economic instruments

ECONOMICS AND THE EU WATER FRAMEWORK DIRECTIVE

For the first time, a European legislative instrument refers to the need to take account of the costs to the environment of undertaking activities in the water field. Article 5 requires an economic analysis of water use in the river basin district to be undertaken. Annex III sets out what must be included. Primarily the detailed examination is aimed towards being able to calculate the costs of water services so as to be able to make use of economic instruments in improving water status in the river basin. Item 38 of the prologue states that the *"use of economic instruments may be appropriate as part of the programme of measures, and that the recovery of the costs of water services including environmental and resource costs associated with damage or negative impact on the environment should be taken into account"* when applying the polluter pays principle. Water or effluent charging regimes and other incentive based options may be valuable aids to controlling water use and polluting discharges, hence leading to reductions in water demand and pollutant load.

Article 9 takes this into the Directive by requiring Member States to take account of the principle of cost recovery including environmental and resource costs. By 2010 water pricing policies have to be introduced that provide incentives to efficient water use, helping to achieve the environmental objectives of Article 4, using as a basis the data obtained in the economic analysis of river basins. However, if Member States consider that they already have adequate systems in place and do not wish to apply new pricing policies, they may choose not to apply this principle provided this *"does not compromise the purposes and achievements of the objectives of this Directive"*. Reasons for not applying the provisions have to be reported in the river basin management plans.

Decisions on the introduction of new regimes for water charging may be influenced by the overall costs associated with the Directive and its new way of managing the environment. According to a report prepared for the Commission prior to the adoption of the Directive (EU 2000) the expected costs will fall into three categories: administrative costs; monitoring costs; and costs involved in achieving the objectives of the Directive.

Administrative costs

These will include the costs of setting up the river basin authorities or other competent body structures to enable river basin districts to operate; the annual operating costs once established, preparing the river basin plans – including the initial survey, establishing objectives, and consultation; monitoring costs, and costs involved in running the authorisation and abstraction registration procedures. The establishment of various public registers will be an added issue in some countries. The implementation of full cost recovery principles will have its own costs. Monitoring costs will constitute a considerable proportion of the new costs, especially in countries that currently undertake limited biological investigation in the water field. The Commission estimated that the costs of ecological monitoring of surface water in the EU would amount to 350 million Euros per annum (mid 1990s), but chemical and quantity aspects could double this figure. Groundwater monitoring would amount to some 30 million Euros. The Commission report estimated the average costs for administration of a river basin authority as 0.5 million Euros (1997 values).

In addition to the costs of simply establishing and maintaining the Directive, there will inevitably be significant costs to improve water bodies to their objective status. Such costs will involve improvements in industrial procedures and wastewater treatment practices, new agricultural practices to reduce diffuse pollution, changes in traffic management, energy production and all other activities of developed nations which could affect water quality and quantity. Many of these improvements are already in the pipeline as a result of existing

EU policies and legislation, for example the introduction of "Best Available Techniques" under the Integrated Pollution Prevention and Control Directive and other environmental protection legislation measures. There is concern over the added costs which could arise as a result of the need to improve water which is "heavily modified" to a good ecological potential. The establishment of new standards for priority substances may also lead to revisions of authorisations for discharges and the additional costs required to meet these.

In the UK the estimated costs of complying with the Directive have ranged from £3.2 billion to £11.2 billion, based on an original full implementation date of 2010 (DETR 1999). The compliance deadlines have extended since this time so costs may be less than this. The costs for administration are estimated at £5–6 million, preparation of plans at £37–54 million, and additional monitoring costs of £144 million up to the year 2040. At present the estimated costs of river monitoring is £32 million per year, and an additional £4.9 million would be needed. Lake monitoring is currently not practised as a large item of monitoring and it would cost an additional £6.4 million per annum to set this up. The total cost of reaching "good status" is estimated at £3 billion to £11 billion. Costs for reducing the impact of point source pollution would amount to £1.4 to £6.5 billion, diffuse pollution £1–3.5 billion. Improvement works to river habitats might cost £100-700 million, low flow alleviation costs £25–250 million and £24 million for other costs. Where is this money to come from? In order to finance this level of expenditure, prices for water services would probably have to rise significantly.

POLLUTER PAYS PRINCIPLE

A fundamental principle in the Directive is the use of the polluter pays principle. Most European nations operate this in respect of charges for water services and pollution control. In many cases the fees for discharge of effluent or for the abstraction and use of water are used to fund the necessary investment in infrastructure, or to pay for the costs of regulation and control. Some country examples are given below.

Portugal

Under legislation introduced in 1994, water charges operate on the user or polluter pays principle. Licensed use of water is subject to a tax (or charge) which is calculated on a proportional basis to the amount used and the economic value of the water to each specific sector, and an inverse proportional factor which takes into account water availability. The law was introduced in 1995 and authorised gradually increasing payments until 1999. The Portuguese system is interesting as money raised by the tax is allocated to improvement in river basin management and implementation of the river basin management plans. There are problems in the enforcement of the tax, partly caused by the need for registration of all water users, and partly as a cultural issue in that abstraction of water was, until this legislation, free of charges. Help was made available to industry and municipalities to introduce the tax. There is a move towards privatisation of the water supply and waste treatment industry that will bring the charges made for water supply up to more realistic values than the heavily subsidised rates previously charged. Metering of water usage is common, and cost to consumers is related to the amount used. Wastewater collection is paid as an additional charge with the water bill, or in some cases as a municipal tax rate.

Netherlands

The costs of maintaining national waterways lies, as might be expected in a country so dependent upon control of water structures, with the government, paid for out of national taxation. Occasionally, harbours and similar structures are financed by the larger municipalities (for example the Rotterdam harbour). The water management tasks carried out by the local water boards may be financed by levies on land owners but they may also receive subsidies from government for these tasks. Groundwater abstraction is subject to a charge. The polluter pays principle operates in the case of wastewater discharges – industry or local inhabitants are subject to a discharge levy mainly under the Pollution of Surface Waters Act 1981.

Germany

In the German example a municipal charge is levied for sewerage services to cover the costs of provision. In practice this is based on water consumption, as most establishments are metered. However, the operation of industrial wastewater treatment plants is the responsibility of industry that bears the cost of construction and operation. Charges for water supply are fixed by the water suppliers. As most of the organisations that supply water are owned by the municipalities, the charges are effectively set by councillors on behalf of the community. Water prices are intended to cover the costs of supply including maintenance and new investment needs. In some cases, where specific requirements exist in addition to the normal needs, the prices to individual users are adjusted to reflect this.

France

The *Agences de l'eau* operate the polluter pays principle by levying charges for abstraction and the discharge of pollutants into watercourses. The income from charges is used to subsidise municipal sewerage, pollution abatement schemes and municipal water supplies. The water abstraction charge is based on the scarcity of the water and how much water is returned to the environment and is related to the vulnerability of the source.

United Kingdom

The water legislation in the UK requires that virtually all the costs of regulation, that is permitting, inspection and sampling of all types of polluting discharges, not simply those which discharge to water, are covered by the fees payable to the Environment Agency. Water and waste water treatment services are provided by privatised companies in England and Wales, and users of the services have to pay an economic charge which covers not only the costs of provision but also provides a profit for the companies concerned. In Scotland such services are provided by public authorities who recover the costs of provision through charges. The "polluter pays principle" operates on water and municipal wastewater treatment in this way. Industry has to bear the costs of services in the same way but, in addition, if there is a need for additional waste water treatment plants, or extra expense to deal with particular pollution problems, industry must bear the total cost of these too.

Luxembourg

Charges for obtaining a permit for discharge into water applies in the Grand Duchy. These charges include a fee for authorisation, inspection and any costs of studies that are needed to ensure that the discharge will have no adverse effects.

ECONOMIC INCENTIVES

Whilst the above describes the costs of implementing the Directive in a technical sense and indicates where such costs may be recovered, the Directive goes further. Article 9(1) requires that environmental and resource costs are taken into account and that water pricing policies provide adequate incentives to use water resources efficiently. This is more than simple cost recovery.

Externalities occur when the activities of an industry give rise to unintentional effects on third parties. For example, the abstraction of water may impose costs on anglers and recreational users of a river because of reduced flows, discharges of pollutants may cause problems to a downstream ecology due to lowered levels of dissolved oxygen. Whilst regulation seeks to minimise such effects, and the costs of regulation is borne by the abstractor or discharger, the costs to the user (or to the ecology) is not taken into account. The Directive seeks to redress this by stating the need to include environmental and resource costs in the equation.

The costs of using the environment, or of causing damage to ecosystems, are not easy to determine and to include within a cost recovery or charging scheme. A number of experimental procedures based on economic principles is described by McMahon and Moran (McMahon and Moran, 2000) to assess the economic value of environmental damage by such techniques as contingent valuation methods, cost-benefit analysis, and hedonic pricing. The techniques are specialised, but are based on such principles as market price approaches, where the environmental change might be assessed as the cost of restoring an asset (say a river or lake) to its previous condition as a measure of its value; contingent valuation, in which an affected population is asked how much value they would place on each option for change of a particular environmental facility (in terms of willingness to pay for its use), and a technique known as hedonic pricing based on changes in property values as a result of environmental degradation. Details of such techniques are beyond the scope of this book.

The OECD has examined incentives for efficient water usage (OECD 1989). These cover five main categories: charges; subsidies; deposit refund systems; market creation; and enforcement incentives.

Charges

Water charges can be set at levels that cover the costs of collection and treatment, so as to induce firms to use water saving devices, or to re-use water. The costs can be set higher than the actual costs. Effluent charges may be set as a price paid for use of the environment – based on pollution load or harm to the environment, or to cover the costs of regulation.

Subsidies

These can include tax incentives, tax credits, grants and low interest loans. The removal of subsidies can act as an incentive to better environmental performance by forcing users to innovate or reduce water use.

Deposit refund systems

Customers pay a surcharge when buying a potentially polluting product. On returning to an approved centre for recycling or disposal their deposit is returned. Such systems can be used for such dangerous goods as pesticide containers or residues.

Market creation

By setting a limit on the total allowable polluting load or abstraction volume users or potential users may indulge in trading of their permitted rights whilst not exceeding the amount of impact that the environment can stand. This system should lead to efficient use of the allowable environmental impact.

Enforcement incentives

These are penalties to induce polluters to comply with environmental standards or regulations. They include fines for exceeding limits, performance bonds-payments to regulatory authorities before a potentially polluting activity is undertaken, which is returned when the correct regulatory levels are met, and liability assignment where polluters are made liable for environmental damage that they cause.

REFERENCES

DETR (1999) Regulatory Impact Assessment for Water Framework Directive, Water Quality Division, DETR, London.

EU (2000) *Use of New Technologies and Cost of Water in View of the New Water Directive of the EU*, Directorate General for Research, Luxembourg.

McMahon P. and Moran D. (2000) *Economic Valuation of Water Resources*, CIWEM, London.

OECD (1989) *Economic Instruments for Environmental Protection*, OECD, Paris.

15

Recording and reporting

REPORTS – DIRECTIVE REQUIREMENTS

The Directive contains many reporting and record keeping requirements. There are two audiences for the reporting obligations – the Commission and the public. The Directive is concerned with involving the public in the management of river basins, and reporting and consulting is a major activity in its achievement, but the Commission also has a responsibility to check that each Member State is obeying the rules, and it has to report to the public itself on this matter. Recording the data obtained in such a way that the public can understand it and so that scientific assessment can be properly carried out is also a crucial duty of the competent authorities The two issues of recording and reporting therefore go hand in hand.

Reporting to the Commission

Reporting to the Commission is formally a responsibility of the government of the Member State, but this may be delegated to a competent authority if this is not a government department. The legal obligation still rests with the government. At every significant stage of implementation the Member States must report progress.

Reports to the Commission are now usually subject to the timescales set out in the Reporting Directive (91/692/EEC) and its later Decision (94/741/EEC). Together these together bring the very varied formats into a common and steady form. The EU Water Framework Directive however sets some specific targets for reporting.

Because the whole emphasis in the Directive is towards the management of water through river basin districts, it is natural that reports of this process are sent to the Commission. Article 15 specifically concerns reporting, and obliges Member States to send copies of their river basin management plans and any updates to the Commission (and other Member States) within 3 months of publication. These include plans for all river basins which fall within their territory and the plans for the parts of international rivers which lie within their area. Reports on the results of analyses and reviews of activities in river basins and the monitoring programmes undertaken to establish the status of waters must also be sent. The first reports will be needed in 2009. The timescale allowed for reviews is 13 years. This means that Member States will be reporting at a wide variety of times during the first 13 years of the operation of the Directive. In addition there is a requirement to report upon progress in carrying out the programme of measures to meet the water status objectives.

Whilst the format of the river basin plan includes details of many of the obligations within the Articles of the Directive, and therefore will form the main report, there are several other obligations amongst the Articles of the Directive for sending details to the Commission of the activities undertaken in connection with its implementation.

Article 3 requires the Member States to inform the Commission of the competent authorities that they have appointed to carry out the terms of the Directive, including the competent authorities of any international body in which they participate. Most countries are signatories to international conventions, so the identification of these would be included alongside the river basin authorities, or the overseeing national body, if that is the chosen route. A very short timescale is involved here. The Commission should have been informed by June 2001. Any changes must also be sent within 3 months from its enactment. Annex I is a detailed list of what is required and this includes the name and address, the geographical coverage of the river basin district, the legal status (including a copy of the document signifying the legal basis), a description of the responsibilities, the membership and any international relationships. Annex I seems to indicate that a competent authority is based within the river basin district, but in practice this will not always be the case as it will be up to the Member States to determine how river basins are managed, and competent authorities may be local, regional or national depending upon the administrative regime in the country concerned.

Article 12 points out that there are many issues that cannot be resolved solely by the Member State. In this case the matter is reported to the Commission which takes a responsibility to solve the problem through consultation with other States.

There is an obligation on Member States under Article 24 to report upon the national laws used to transpose the Directive.

Reporting to the public

Consultation is an important element. Article 14 sets out the formal requirements for public information and consultation. This Article concerns the river basin plans, and all consultation is made in relation to it. The reporting programme consists of a timetable and programme for the river basin plan production, an overview document for the river basin and draft copies of the plan. If there are any background documents these must also be made available to the public and others who may be affected. Under the terms of the Freedom of Access to Environmental Information Directive (90/313/EEC), all of the information connected with the formulation of the plan, and all of the subsequent monitoring data and reports derived from these must also be made available, although this is not expressly mentioned as an EU Water Framework Directive obligation. This means that an adequate and easily accessible system must be set up in time for the first of the information to be made available by 2008, one year before the final river basin plan is published. However, account must be taken of the need for much earlier reporting, particularly in respect of the assessment undertaken for Article 5 (review of the characteristics of the river basin). Article 14 suggests that *all interested parties* should be involved at in the production review and updating of the river basin plans. In practice, as it will be necessary to determine what activities take place in the catchments and how water is used, it will be impossible to produce a plan without the active involvement of water users and people who live and work in the river basins, so Member States will have to establish adequate reporting and consultation mechanisms from the start.

Many countries already operate public registers of environmental information, some of which, like the UK example, are legal requirements under internal legislation (UK 1991). Some of the current EU directives have reporting specifications which are designed to allow for public access to the results of their implementation (for example the reporting maps of the Freshwater Fish Directive and the Bathing Water Directive).

Reports produced by the Commission

In addition to reports going from Member States to the Commission, there are heavy obligations for the Commission itself to produce reports for Parliament or the public. To coincide with the river basin planning cycle, the Commission has to use the data submitted to it to produce reports starting 12 years after enactment of the Directive, that is, in 2012; at 6 yearly intervals, on the implementation of the Directive; and also every 3 years from a starting date of 2012, an interim report on progress. The interim reports are based on the reports from Member States submitted under Article 15(3), which themselves are interim reports. All of these reports have to be submitted to the European Parliament, and they will thus become public documents. Further involvement of interested persons (NGOs, consumer bodies and experts) must be ensured by the convening of conferences *when appropriate in line with the reporting cycle.* This is to enable comments and the sharing experience of the Directive.

Two years after receipt of the reports from Member States concerning their monitoring programmes and the investigations into the characteristics of river basins which are used for their first river basins plans, the Commission must itself prepare a report, assembled from the information in these summary reports. As the work to characterise river basins must be established four years from the date of publication of the Directive, the Commission's reports are required in 2006 and 2008. Furthermore, at yearly intervals the Commission also has to prepare a plan indicating what measures it is considering, which may have an impact upon water legislation. This report will have to be presented to the regulatory committee convened under Article 21.

REPORTS – FORMAT

The precise form of reports both from Member States to the Commission and by the Commission to the Parliament is not specified except in some particular aspects. As most of the reports are based on the content of river basin plans, it must be assumed that the reports will have to contain the details required under Annex VII – the elements comprising the river basin plans. Each element requires certain information and the details of this will be affected by how the Member States set out their plans, and on the nature of the problems in the river basins. However, there are rules laid down for presentation of environmental status. Section 1.4 of Annex V1 describes the presentational requirements. Most countries in Europe (and indeed elsewhere in the world) use individually designed presentational systems for showing the quality of their surface and groundwaters. These are often based on a mapping system using colour codes often supplemented by other symbols. The Freshwater Fish Directive, for example used a colour-coded map with additional symbols representing particular problems. Some countries use maps to display bathing water quality, and many prepare annual reports of various descriptions. In order to provide a consistent and harmonised view of environmental status, the Commission has set up a common methodology for the preparation of maps under the Directive.

Presentation of monitoring results and status

The Directive aims to achieve good status in surface and ground waters. For surface waters the ecological and chemical status must both be reported in the form of maps covering the river basin district. In the case of groundwater, maps showing the quantitative and chemical status must be prepared. The maps will form part of the river basin plan documents. The Directive specifies a system of colour codes for each of the status assessments.

Ecological status of surface waters

Ecological status is assessed against the values for biological and physico-chemical parameters, and the status is equivalent to the lower of the two possible values. On the map the data must be presented using the colour codes:

Blue	=	high status
Green	=	good status
Yellow	=	moderate status
Orange	=	poor status
Red	=	bad status

A variation on these colours is used for showing ecological potential in artificial and heavily modified water bodies. The relevant colours are as above but with the insertion of light grey and dark grey stripes into the code. There is no separate code for showing high ecological quality, however, the green colour code (equal green and light grey stripes for artificial water bodies and green and dark grey stripes for heavily modified water bodies) shows good quality and above. There is also a facility on the maps for showing bodies of water that fail to meet environmental quality standards for synthetic or non-synthetic pollutants and thereby missing the status of good ecological quality. This is shown by a black dot.

Chemical status of surface waters

Chemical status is shown by a simpler system. Waters which meet all the environmental quality standards of Annex IX, priority substances in Article 16 and other EU legal standards, are classified as good chemical status and coloured blue on the map. If there are failures to achieve these levels the colour code is red. There seems to be some potential confusion here between the use of blue and green colours.

Quantitative status of groundwaters

Good quantitative status is related to the groundwater level regime, and if the available groundwater is not exceeded by the long-term annual average rate of abstraction the groundwater is of good quantitative status shown by a green colouration on the map. If abstraction exceeds recharge, the water is of poor quantitative status, shown in red.

Chemical status of groundwaters

In assessing the chemical status of groundwater, the results from individual monitoring points are aggregated for the groundwater body as a whole and shown as green for good status and red for poor status. The status is assessed against the requirement for good status, which means that the water must show no saline intrusion, comply with the quality standards of Community groundwater legislation, and not have an adverse effect on any associated surface waters nor on any dependent ecosystem. Especially in the case of groundwaters, if there is any significant upwards trend in pollutants as a result of human activity, this must be shown on the map by a black dot. The Directive allows Member States to combine the two groundwater maps provided the trends are indicated.

Implications

Member States will have to adjust their varied mapping procedures to comply with the new harmonised system by the time the first draft river basin management plans are prepared for public consultation, that is, by 2008. The final form of maps will be required to accompany the report to the Commission in 2009 of the first set of river basin plans for their territories.

REFERENCE

UK (1991) *The Environmental Protection (Applications Appeals and Registers) Regulations*, Statutory Instrument 507.

16

Summary timescales and future developments

SUMMARY TIMESCALE

This implementation of this Directive is spread over a much longer period than most. This reflects the great complexity and the acknowledgment that many of the tasks are of themselves long-term issues. For example, the necessary survey work to formulate a river basin management plan will take time. The completion of programmes of measures must allow for large constructional work or major changes in the way that industry and agriculture carries out it work. The response of biological organisms to changes in their environment is unpredictable and may take many years. To see real improvements in some waters may take in excess of 20 years. A picture of the overall position in terms of dates in the Directive is shown in Table 16.1.

Table 16.1 Overall timetable for implementation of the Directive.

Activity	Key date
Transpose Directive into national legislation	2003
Define river basins, appoint competent authorities	2003
Complete surveys	2004
Commence monitoring programmes	2006
Statement of issues	2007
Publish draft river basin plans for consultation	2008
Commence river basin plans	2009
Enact programme of measures	2009
Introduce water pricing	2010
Implement all programmed measures	2012
Achieve good water status in most waters	2015
First review of river basin plans	2015
Second review of river basin plans	2021
Where extensions apply achieve good water status	2027
Third review of river basin plans	2027

REPEALS

The Framework Directive will replace a number of existing legislative instruments and will change others. The repeal provisions are phased to match the enactment and implementation of parts of the new Directive so that there will be a seamless transition from the old to the new. According to Article 22, repeals of three Directives will take place in 2007. These are the Directive concerning the quality of surface water intended for abstraction of drinking waters (75/440/EEC) and its associated Directive concerning methods of measurement (79/869/EEC) and the Directive on the exchange of information on the quality of surface freshwater (77/797/EEC). These Directives will be effectively superseded at an early stage in the implementation programme by the monitoring programmes set up under Article 8, which should be established by 2006 and the identification of waters used for the abstraction of drinking water through Article 7. The provisions of the revised Drinking Water Directive of 1998 (98/83/EC) should have been transposed by 5 December 2000, and the appropriate measures taken to meet those drinking water quality standards by 2005 (except where there are extended timescales or exceptional circumstances). Article 4 of the new Drinking Water Directive requires Member States to apply appropriate types of water treatment to waters to ensure that the quality standards are met, and this supersedes the need for the classification system adopted in 75/440/EEC. The new monitoring programmes absorb the provisions of the other two repealed Directives.

As the river basin plans become established, the need for several other directives is lessened. The river basin plans will indicate quality objectives for those waters that require protection for freshwater fish and shellfish, and the two directives concerning these types of water – the Freshwater Fish Directive (78/659/EEC) and the Shellfish Waters Directive (79/923/EEC) are repealed as the river basin plans become operational in 2013. At the same time the Groundwater Directive (80/68/EEC) is repealed, as groundwater protection becomes an issue in the plans.

As explained in Chapter 12, a new system for dealing with dangerous substances is introduced by this Directive, and as a consequence, the Dangerous Substances Directive is also repealed at the same time, although, Article 6 of that Directive is repealed immediately. This is the Article that obliges the Commission to lay down limit values and quality objectives for dangerous substances. Chapter 12 sets out the new procedure which covers this issue from the date that the Framework Directive became operational. There are some transitional provisions in respect of the Dangerous Substance Directive. The list of priority substances in the new Directive replaces the original list of dangerous substances; and until it is repealed, the principles of Article 7 of the Dangerous Substances Directive are to be carried out by using the provisions of the new Directive in terms of defining problem substances, and setting quality standards and adopting programmes of measures. This means that toxicity assessments may be used to determine appropriate standards. The Dangerous Substances Directive will continue in all but name, however, as the first river basin management plan must establish quality standards which are *at least as stringent* as those required by that Directive. Although some of the existing substances will not appear on the priority lists, it seems that they will continue to be legal standards once quality standards have been set, as it is not possible to permit a deterioration in quality under this Directive.

FUTURE PROPOSALS

The Commission has many jobs to do under this Directive, which will lead to possible changes. It is assisted by a Regulatory Committee, set up by virtue of Article 21 which will lay down the rules of procedure.

The Committee will have to be established rapidly, as within two years from the date of publication in the Official Journal the Commission has to present an indication (*an indicative plan*) to the regulatory committee of the measures which it intends to develop. Article 16 gives the Commission a specific task in relation to priority substances, as explained in Chapter 12, and, as it will be difficult to fulfil this obligation within the timescale set, the plan is likely to be an outline only.

Nineteen years after publication, the Commission has to review the Directive and propose any necessary changes. This period will allow the first of the river basin plans to have been completed, and the success or otherwise of the policies in reaching "good water status" in the Member States will be apparent.

Index